The Physics of Non-Thermal Radio Sources

NATO ADVANCED STUDY INSTITUTES SERIES

*Proceedings of the Advanced Study Institute Programme, which aims
at the dissemination of advanced knowledge and
the formation of contacts among scientists from different countries*

The series is published by an international board of publishers in conjunction
with NATO Scientific Affairs Division

A	Life Sciences	Plenum Publishing Corporation
B	Physics	London and New York
C	Mathematical and Physical Sciences	D. Reidel Publishing Company Dordrecht and Boston
D	Behavioral and Social Sciences	Sijthoff International Publishing Company Leiden
E	Applied Sciences	Noordhoff International Publishing Leiden

Series C – Mathematical and Physical Sciences

Volume 28 – The Physics of Non-Thermal Radio Sources

The Physics of
Non-Thermal Radio Sources

*Proceedings of the NATO Advanced Study Institute
held in Urbino, Italy, June 29 – July 13, 1975*

edited by

GIANCARLO SETTI

*University of Bologna, Laboratorio di Radioastronomia CNR,
Via Irnerio 46, 40126 Bologna, Italy*

D. Reidel Publishing Company

Dordrecht-Holland / Boston-U.S.A.

Published in cooperation with NATO Scientific Affairs Division

ISBN-13: 978-94-010-1519-6   e-ISBN-13: 978-94-010-1517-2
DOI: 10.1007/978-94-010-1517-2

Published by D. Reidel Publishing Company
P.O. Box 17, Dordrecht, Holland

Sold and distributed in the U.S.A., Canada, and Mexico
by D. Reidel Publishing Company, Inc.
Lincoln Building, 160 Old Derby Street, Hingham, Mass. 02043, U.S.A.

CONTENTS

FOREWORD

This volume contains a series of lectures presented at the 3rd
Course of the International School of Astrophysics held in Urbino,
Italy from June 29 - July 13, 1975 under the auspices of the
"E. Majorana" Centre for Scientific Culture. The course was jointly
planned by L. Woltjer and myself and was fully supported by a grant
from the NATO Advanced Study Institute Programme. It was organized
with the aim of providing students and young researchers with an
up-to-date account on the subject of the structure, origin, and
evolution of non-thermal radio sources and was attended by 88
participants from 15 countries.

It is well known that radio astronomical research has played
a fundamental role in the development of modern astrophysics and
cosmology, leading in particular to the exciting discoveries of
new and unsuspected classes of objects such as radio galaxies, quasars and pulsars. However, it is probably fair to say that we are
still far from a complete understanding of the physics of non-
thermal radio sources except for the fact that, in general, the
radio emission is adequately explained in terms of the "synchrotron" mechanism.

In the case of extragalactic sources, we know that the very
large amounts of energy observed are supplied as a consequence
of violent activity taking place in galactic nuclei. Yet we are
still faced with a number of unsolved problems, for example: what
is the nature of the ultimate source of this energy, how and where
is the energy so efficiently converted into relativistic particles,
and how is the energy collimated to millions of light years away
with such a high degree of symmetry, as found in the double radio
sources.

Phenomena observed in extragalactic sources reoccur with
striking similarities in galactic non-thermal radio sources, though,
of course, on different space and time scales. For instance, the
radio properties of some X-ray stars like Sco X-1 and Cyg X-3 are
typical of the phenomenology found in quasars. Moreover, the discovery of the association between supernova remnants and pulsars
has triggered the notion that very efficient particle acceleration
can be obtained by rotating magnetic objects.

On the observational side, a big step forward has been obtained by the completion of large aperture synthesis radio telescopes, which have permitted a detailed mapping of radio sources, and by the development of very long base-line interferometry (VLBI), with resolutions down to 10^{-3} - 10^{-4} arc sec, which for the first time has rendered possible radio investigations of the structures of the very compact radio sources found in galactic nuclei and quasars. The observations obtained with this new instrumentation, combined with the work in the optical, infrared and X-ray domain, have made great strides in recent years toward improved testing of radio source models.

We feel that the material presented at the Institute represents a rather complete and comprehensive coverage of the present status of studies of non-thermal radio sources, though one is aware of rapid developments in this field of research. The various aspects of this exciting subject were covered in a series of lectures, presented in this volume, totaling 44 hours and in 14 topical seminars given by the participants.

I wish to express my gratitude to the Scientific Affairs Division of the North Atlantic Treaty Organization for the generous support given to the course. Sincere thanks are also due to Mr. L. Baldeschi for helping with the organization of the meeting and for drawing a number of figures contained in this volume; to Mrs. B. Mandel for the patient typing and help in the editing; and to Mr. R. Primavera for the photographic reproduction of part of the figures.

A special thanks is, of course, due to all the lectures and participants who contributed so much to the success of the course and provided the articles for this volume.

Finally, I wish to express my sincere gratitude to L. Woltjer for his fine collaboration as co-director of the course.

Prof. Giancarlo Setti
University of Bologna
Laboratorio di Radio-
    astronomia CNR
Via Irnerio 46
40126 Bologna, Italy

MORPHOLOGY OF EXTRAGALACTIC RADIO SOURCES

George Miley

Sterrewacht, Huygens Laboratorium
Leiden, The Netherlands

1. TECHNIQUES FOR MEASURING MORPHOLOGY

"The astronomical instruments have been in use from very ancient days, handed down from one dynasty to another, and closely guarded by the official astronomers. Scholars have therefore had little opportunity to examine them, and this is the reason why unorthodox cosmological theories were able to spread and flourish." - Fang, Hsüan-Ling, History of the Chin Dynasty, 635 AD.
This section is intended to be a 'bluffers guide' for readers of morphology papers.

1.1 Some peripheral techniques

a. Single dish

Resolution $\Delta\theta \sim \lambda/r$ (for r = 100m (Bonn dish) at λ = 3 cm, $\Delta\theta \sim 1'$). Only useful for largest sources at short wavelength. Observations seriously limited by receiver stability.

b. Lunar occultations

Variation of source flux density as it is occulted by moon gives one-dimensional brightness distribution convolved with diffraction pattern of moon. For restoration procedure see Scheuer (1962), von Hoerner (1964). $\Delta\theta$ depends on signal to noise of source. $\Delta\theta \sim 1"$ readily achievable.

G. Setti (ed.), The Physics of Non-Thermal Radio Sources, 1–26. All Rights Reserved.
Copyright © 1976 by D. Reidel Publishing Company, Dordrecht-Holland.

Advantages   : Resolution not highly dependent on frequency.
Disadvantages: Two-dimensional structure difficult to reconstruct.
Can only be applied to sources that happen to lie on moon's path
(i.e. $- 29° < \delta < + 29°$).
First suggested by Getmansev and Ginzbury (1950). Most notable
success in pinpointing 3C 273 (Hazard et al. 1963) led directly
to discovery of quasars. Recently several hundred weak sources
have been studied with a new telescope in India specially designed
for occultation work, e.g. Kapahi et al. (1974).

c. Interplanetary scintillations

Irregularities in electron density of solar corona (scale $\sim$ 100
km) cause small sources to flicker or "scintillate". Large sources
are unaffected - frosted glass effect. Degree of scintillation
and its variation with distance from Sun can give useful information on amount of small scale (< 1") structure in a source. First
reported by Hewish et al. (1964). See Cohen et al. (1967), Little
and Hewish (1968).
Advantages   : Easy. High resolution at low frequency.
Disadvantages: All but crudest inferences about structure depend
on assumed model for solar corona. Can only be applied to limited
source list ($\sim$ 50° from Sun).

1.2 Interferometry and aperture synthesis

This is much the most important tool used by morphologians. It
was first developed by Ryle (1952) and independently by Christiansen
(1953) and Mills and Little (1953). An excellent account is given
by Fomalont and Wright (1974). Here we shall summarize some of
the most important points stressing the jargon which is often
used by morphologians in their articles

Analogue of Young's
double slit experiment.

Gives roughly same
resolution as single dish
stretching between two
telescopes.

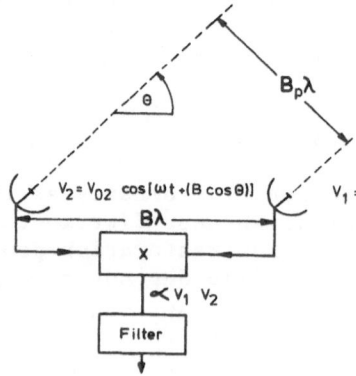

For point source of flux density S the output is

$$S_e(\theta) = S \cos (2\pi B \cos \theta) \qquad (1.1)$$

As earth rotates θ changes. Output is a series of sinusoidal
interference fringes. Positive and negative fringes correspond
to different θ, i.e. different positions in the sky. The *interferometer polar diagram* can be pictured as a comb-like structure
of positive and negative lobes in the sky. This depends only on
the positions of the two telescopes, not on their directional
properties. The separation between the two telescopes is called
the *baseline* B. The *projected baseline* vector B_p, defines the
resolution of the interferometer.
Angular distance between two positive lobes $\Delta\theta \sim 1/B_p$ radians.
A source of angular size $\Delta\theta \sim 1/(2\ B_p)$ occupies positive and negative lobes simultaneously. The resultant fringes are blurred,
i.e. reduced in amplitude and changed in phase. A study of this
fringe blurring as a function of projected baseline clearly gives
useful information about source structure.
How does the projected baseline change as the source moves through
the sky? At any instant B_p can be resolved into an east-west component, u, and a north-south component, v. As the earth rotates
the tip of B_p describes an ellipse in the *u-v plane* (i.e. the
trajectory eccentricity of this *resolution ellipse* is cos δ, i.e.
depends on the declination of the source. For $\delta = 0$ it becomes
a straight line. For $\delta = 90°$, a circle. Note that for an *E-W
interferometer* resolution ellipses are always centered on origin.
In this case *for source at equator ($\delta = 0$) no N-S resolution can
be obtained.*
It is well known and easily shown that the amplitude I and the
phase ϕ of the interferometer fringes are the amplitude and phase
of the Fourier transform of the brightness distribution on the sky,
I (1, m). Thus

$$I(u,\ v) e^{i\phi}(u,\ v) = \int_{-\infty}^{\infty} I(1,\ m)\ e^{2\pi i(ul\ +\ vm)}\ dl\ dm \qquad (1.2)$$

where (u,v) gives the instantaneous projected baseline and ϕ(u,v)
is the phase measured with respect to a point source at 1 = 0,
m = 0.
If I and ϕ are measured at *all* projected baselines out to ($u_m\lambda$, $v_m\lambda$),
Fourier inversion gives a map of I(1, m) witn a resolution in
radians $\sim 1/u_m$ (east-west) and $1/v_m$ (north-south). This application
of interferometry is called *"earth rotational aperture synthesis"*
because it gives a similar resolution to a filled aperture which
covers the baseline.
Similar Fourier relations exist for the other 3 Stokes parameters,
Q, U, V which represent the linear and circular polarization in
the source. By having two orthogonal linear or circularly polarized
receivers on each telescope and by combining or correlating the
4 pairs, I-fringes, Q-fringes, U-fringes and V-fringes can be

measured simultaneously. Interferometers make excellent polari-
meters and therefore can be used to map the polarization distribution in sources.

1.3 Some jargon associated with aperture synthesis

a. Map distortions

An aperature synthesis map is distorted by ripples or *sidelobes*
due to incomplete baseline coverage. Mathematically they can
usually be expressed in terms of Bessel functions. Most important
are:
(i) *near—in sidelobes* due to abrupt cut off at edge of array.
Radii = $(1/u_m, 1/v_m)$. Most severe close to source. Can be mini-
mized (at cost of resolution) by giving data from longer baselines
relatively less weight, i.e. *tapering* the data with *a grading
function* before Fourier transformation.
(ii) *diffraction grating rings* produced by discrete nature of
array. For a regularly spaced E-W array with adjacent elements
spaced by Δu wavelengths, the grating rings are ellipses centered
on the source with radii $\Delta l = 1/\Delta u$, $\Delta m = 1/(\Delta u \sin \delta)$ radians.
(iii) effects due to unsampled short baselines. Sometimes (e.g.
because it is physically impossible) the shortest spacings of an
otherwise filled comb are unavailable.
Missing zero spacing $\to$ constant negative offset.
Other missing short spacings $\to$ variable negative offset of "bowl".

b. Field of view

Usually limited by the directional properties of the individual
antennas in the array, i.e. attenuation of the *primary beam*.
Another limitation is set by the size of array that is transformed.
To save computing time the Fourier transform is often carried out
using the Fast Fourier Transform (FFT) algorithm. In this case
the data must be convolved to a uniformly spaced rectangular grid
of points before transformation. The effect of this *gridding* or
rectangular sampling of the data in the (u-v) plane is that
responses of the instrument outside the required field of view
are *aliased* or reflected into the picture, as shown in Fig. 1.

c. Synthesized beam

Calculated response to point source observed with exactly same
baseline coverage and transformed in exactly the same manner.
Includes effects of resolution and map distortions. Distinguish
between 'synthesized beam' and 'primary beam'.

d. Source subtraction and restoration

To observe weak structure it is often necessary to remove the

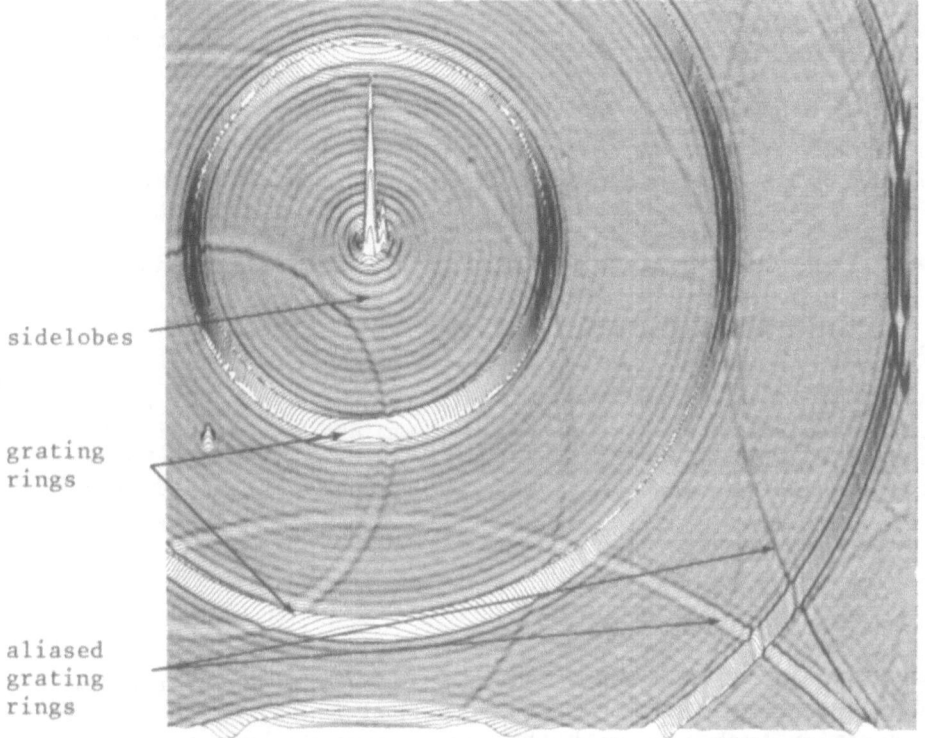

sidelobes

grating
rings

aliased
grating
rings

Fig. 1.   Portion of a typical synthesis map.   Top and left cut off.

ripples and other effects due to stronger sources in the field.
Subtraction including the effects of gridding is expensive but
can be carried out for point sources.   Extended sources are usually
subtracted using a
map *cleaning* technique
which does not
take gridding into
account and thus
does not remove the
effects of aliasing.
'CLEAN', developed by
Högbom (1973) is an
iterative process which
approximates an extend-
ed source by a number
of point sources each
of which is centered

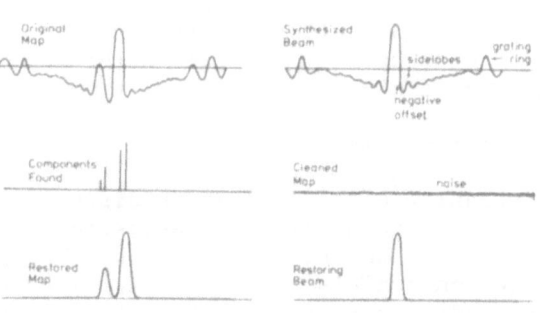

at one of the grid points of the map array.   Starting with the
maximum intensity, the synthesized beam is used to subtract the
appropriate point source response from the complete map.   This
procedure is continued until the intensity remaining in the
region of interest is less than a specified value.   The result is

(i) a cleaned map with the response of the strong source removed
and
(ii) a series of (delta function) intensity points.
The intensity points determine the brightness distribution of the
source and can be *restored* using an artificially defined clean
synthesized beam. The restored map gives the source brightness
distribution without distortions due to sidelobes, grating rings,
etc.
Caution. It is difficult to estimate uncertainties introduced
into the map by the clean restoring procedure.

e. Dynamic range

Response of the instrument to an actual point source will differ
from that predicted by the synthesized beam due to instabilities
in instrumental gain, phase, geometry, etc. This limits the extent to which the effects of strong sources can be removed from
the map. The ratio of the maximum intensity to the minimum intensity that can be usefully studied is called the *dynamic range*
of the map.

f. Spectral comparisons

Difficult due to different baseline (in number of λ) coverage in
different frequency maps. Preferable method is to use the cleaning restoration technique with the same restoring beam at each
frequency. Take care and be skeptical!

2. APPLICATIONS OF MORPHOLOGY STUDIES

"'The cause of lightening' Alice said very decidedly, for she felt
quite certain about this 'is the thunder - no, no!' she hastily
corrected herself. 'I meant the other way'." - Lewis Carroll,
Through the Looking Glass.

2.1 Physical parameters

Total intensity and polarization distributions determine several
physical parameters which characterize conditions both internal
and external to the source. Thus morphology studies can:
(i) place constraints on models of radio sources.
(ii) use radio sources as probes of their environment.
A summary of most of the important parameters that can be derived
in this way are given in Figs. 2 and 3. For an account of the
detailed derivation methods, see Pacholczyk (1970) and the lecture
by Scheuer. Note the multitude of assumptions (italics) necessary
to derive physical parameters (dashed boxes) from observed parameters (solid boxes).

(a) INTENSITY DISTRIBUTIONS

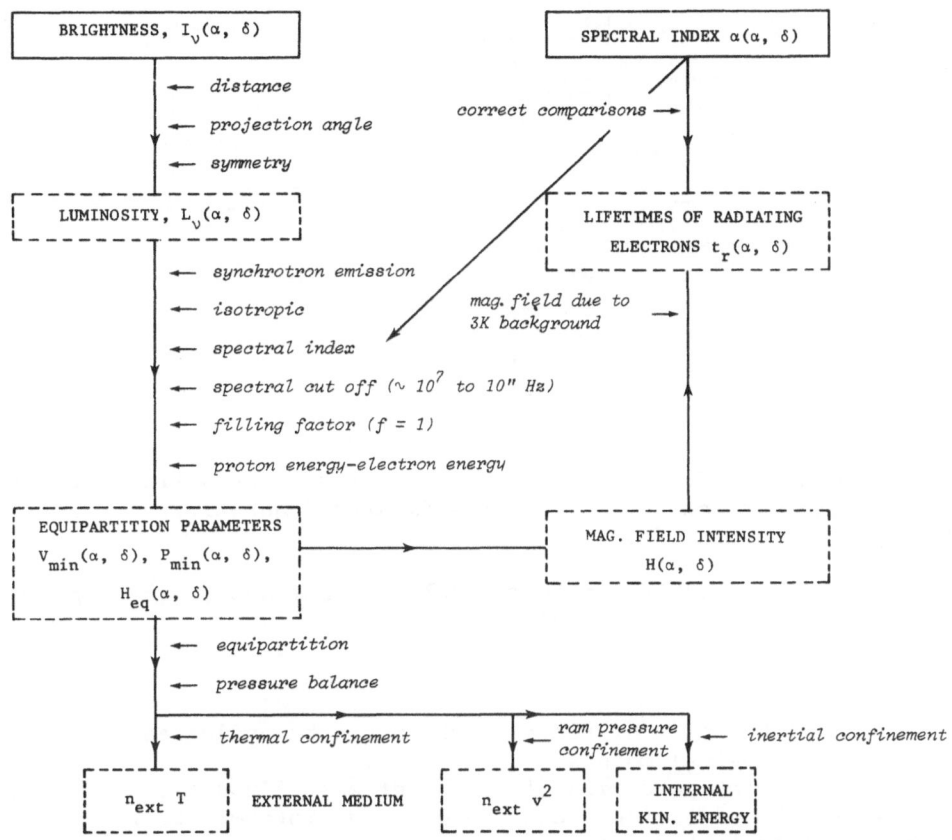

Fig. 2. Schematic diagram of the most important parameters of radio sources derivable from the intensity distribution measurements. Solid boxes contain observed quantities and dashed boxes contain derived quantities.

2.2 Morphology as a cosmological probe

By comparing the structures of near and distant radio sources we can clearly obtain information about the evolution of the universe. So far this has only been done very crudely using the angular size since not enough is yet known about the detailed brightness and polarization distributions of the more distant sources.

a. Angular size vs. redshift

The apparent angular size θ of an object of linear size L depends on its distance and on the geometry of the universe (e.g. McVittie 1965a).

(b) POLARIZATION DISTRIBUTIONS

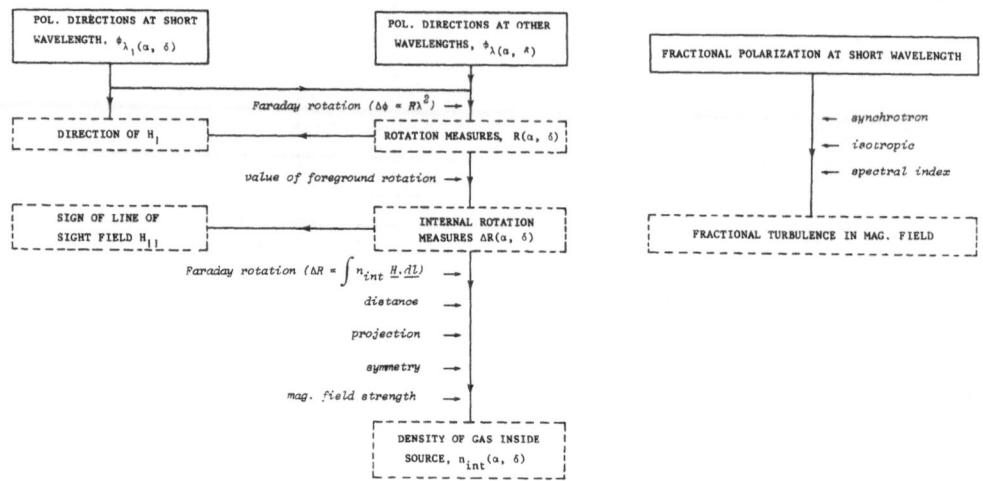

Fig. 3. Schematic diagram of the parameters derivable from the polarization measurements. Solid boxes (measured quantities), dashed
boxes (derived quantities).

*Assuming that the redshift z is a Doppler effect due to the expan-
sion of the universe* we can write

$$\theta = L \ C_m(z)$$

where $C_m(z)$ is function of redshift which can be evaluated for
different cosmological models.
If one has a decent yardstick (i.e. object with fixed L), the study
of the variation of θ with z can yield useful information regarding the geometry of the universe.
Zero pressure cosmological models can be characterized by a deceleration parameter q_o and a density parameter σ_o. The function
$C_m(z)$ has been tabulated for various models (different q_o, σ_o) by
McVittie (1965b) and Refsdal et al. (1967). For all simple models
a minimum occurs in the θ-z relation. In the limiting case (steady
state cosmology) the minimum is approached assymptotically as $z \to \infty$.

b. The θ-z relation for radio sources (Fig. 4).

Individual measurements contain projection effects. Note the behaviour of the upper envelope in Fig. 4. Conclusions from present
data;
(i)   Largest angular size (separation of double) decreases with red-
      shift.
(ii) Component size decreases with redshift.
(iii) There is continuity in behaviour between radio galaxies and

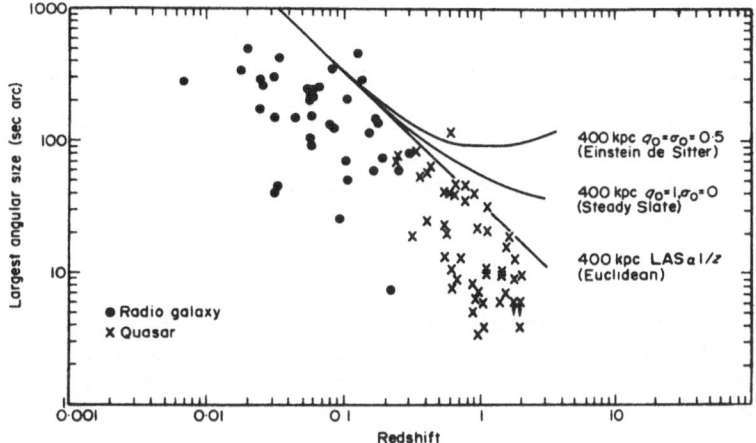

Fig. 4. Diagram from Miley (1971) for steep spectrum sources (with α < -0.6). See also Wardle and Miley (1974).

    quasars.

(iv) Decrease of θ with z is faster than predicted by simple cosmological models.

Possible explanations of faster decrease of θ with z:

(i)   Selection effects in sample.
     The sample was not homogeneous. However, if the results were contaminated by a flux density - redshift relation one might expect intrinsically weaker sources to lie close to the envelope. This is not the case.

(ii) Physical effects at large z, e.g. a denser intergalactic medium could impede motion of radio plasmoids.
     Another possibility is that at large z, the 3K background radiation + inverse Compton process scatter relativistic electrons to lower energies. They radiate X-rays and the radio sources are prematurely snuffed out.

(iii) Cosmological effects.
     We know that the universe is not a homogeneous fluid as required by conventional cosmological theories. It contains galaxies, clusters, etc. Roeder (1975) has shown that inhomogeneities will push the minimum in θ to larger z.

(iv) Origin of the redshifts.
     If z has a non-cosmological component the θ-z relation gives no relevant cosmological information.
     Hewish et al. (1974) from scintillation measurements claim to see a minimum in the θ-z relation. However, their evidence is unconvincing in view of the tiny sample involved and the uncertainties of interpreting scintillation data.
     Further work is needed before any definite conclusions can be drawn.

c. Angular size vs. flux density

Recently workers at Ooty (Swarup, 1975; Kapahi, 1975) have shown
that a useful correlation exists between angular size and flux
density. A similar effect has been found in the Westerbork data.
The angular size flux density relation and its applications to
cosmology will be discussed in a seminar by Ekers.

3. OBSERVED MORPHOLOGY

"Man's knowledge of matter is knowledge of its forms of motion,
because there is nothing in this world except matter in motion
and this motion must assume certain forms." - Mao Tse Tung, On
Contradiction, 1937.

3.1 Introduction

Here we shall summarize the various structures observed in strong
extragalactic radio sources and explain how the observed morphologies place several restrictions on source models. We shall deal
only with strong sources which are well resolved by existing synthesis arrays (such as the One Mile and Five Kilometer telescopes
at Cambridge and the WSRT at Westerbork) (i.e. > 10") and usually
identified with elliptical galaxies or QSOs. Excluded are:
(i)   More compact sources. (See lectures of Kellermann.)
(ii) Sources associated with spiral galaxies (e.g. see Ekers, 1975).
(iii)The extended sources covering some clusters of galaxies.
      (Little is known about the properties of these enigmatic
      sources except that they have very steep non-thermal spectra.)
We shall be concerned mainly with the structure of 3C sources because the 3C survey is still by far the most studied sample of
sources. Note that surveys such as 3C are complete to a given
observed flux density at a given observing frequency over an in-
strumentally determined range of angular size. The properties
of sources in such a survey are therefore not necessarily typical
of those from a sample complete to a given *intrinsic* luminosity
(see the lecture of Ekers).

3.2 Morphological Classification

Morphological classification schemes are somewhat arbitrary and
subjective. It would be a useful exercise if before reading further the student devises his own classification scheme for radio
structures using the following (not comprehensive) list of references to synthesis maps:
(i)   Cambridge One Mile:
        Ryle and Windram (1968), Macdonald et al. (1968), Mackay
        (1969), Elsmore and Mackay (1969), Mitton and Ryle (1969),
        Graham (1970), Branson et al. (1972), Riley (1972),

Harris (1972), Riley and Branson (1973).
(ii)   Cambridge Five Kilometer:
Hargrave and Ryle (1974), Pooley and Henbest (1974), Turland
(1975).
(iii) Westerbork (WSRT):
Miley and van der Laan (1973), Miley (1973), Högbom and
Carlsson (1974), Willis et al. (1974), Miley et al. (1975).
(iv)   Cal Tech:
Fomalont (1971).
(v)   NRAO:
Hogg (1969), Hogg et al. (1969), Macdonald and Miley (1972),
Wardle and Miley (1974).
(vi)   Southern sources:
Cooper et al. (1965), Ekers (1969), Schlizzi and McAdam
(1975).
Table 1 gives a possible classification scheme for radio structures.
We shall treat each of those types of sources in turn.

3.3 Collimated doubles

Fanaroff and Riley (1974) have shown that these sources have
$P_{178} > 2 \times 10^{25}$ W Hz^{-1} ster^{-1} and are on average one to two orders
of magnitude more luminous than any of the other classes of sources
listed in Table 1.

Table 1

Structure of Strong Radio Sources

Type	Example	Fraction in 3C
Collimated double[†]	Cygnus A	70%
Relaxed double[†]	Fornax A	10%
Head tail	3C 129	10%
Complex	3C 84	10%

[†]Doubles is somewhat of a (historical) misnomer since as we shall
see most 'double' sources have a third component between the two
dominant lobes.

a. Main characteristics:

(i)   two collimated extended lobes A,
       A' symmetrical about visible
       object
(ii)   'hot spots' B, B' at outer edge
       of components
(iii) compact nuclear component (core)
       C coincident with nucleus of
       galaxy or QSO

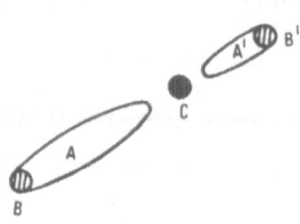

b. Most information from:

(i)   3C 236, the largest angular and linear radio galaxy (Fig. 5)
(ii)   Cygnus A, the closest of the strongest radio galaxies
       (Fig. 6)

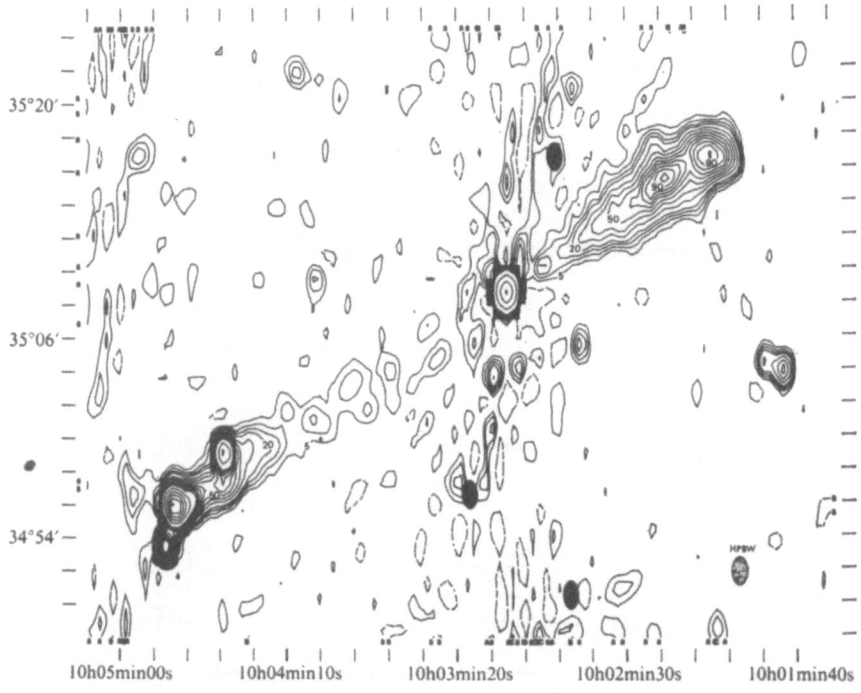

Fig. 5. Contour map of 3C 236 done with the Westerbork Synthesis
Radio Telescope at 49 cm (from Willis et al. 1974). Stripes extending approximately north-south from the central intense source
are due to instrumental effects as discussed in the text. The
three black ellipses represent intense background sources; the
effects of which have been removed from the map.

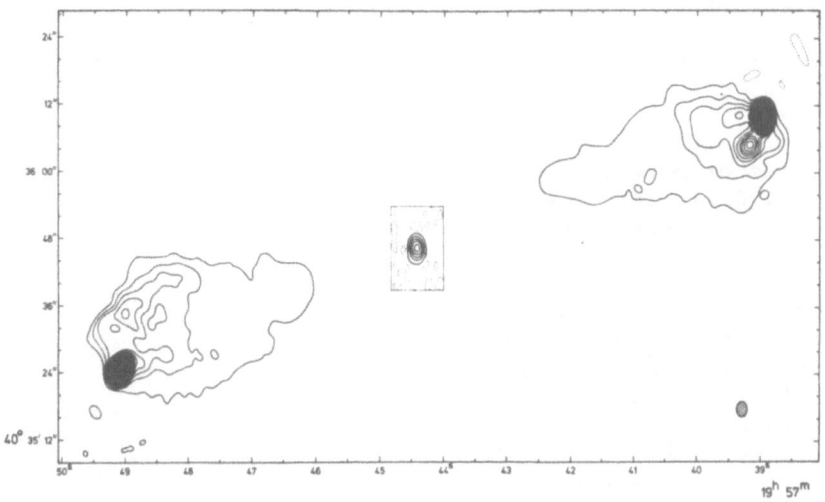

Fig. 6. Cygnus A at 5 GHz with an angular resolution of ∿ 2"
(from Hargrave and Ryle (1974)). The 'hot spots' (blacked out
in diagram) are unresolved at 5 GHz.

c. Properties of extended lobes:

(i)   Steep non-thermal spectra (e.g. α = -1.2 for Cyg A).
(ii) Large percentage polarizations (to 25% in Cyg A and larger
in 3C 236), i.e. ordered magnetic fields.
(iii) Integrated polarization p.a.s. indicate that E is perpendicular to source axis, i.e. magnetic field H is aligned along
axis.

This effect, originally pointed
out by Gardner and Whiteoak (1964)
was found to be less convincing by
later workers (e.g. Mitton, 1972).
I contend that the effect diminished because more sensitive receivers enabled polarization measurements to be extended to less
highly polarized sources where
the field is less ordered. If,
instead of lumping all sources
together, one takes only sources
with *large* percentage polarizations (ordered magnetic fields)
the effect is again very strong.

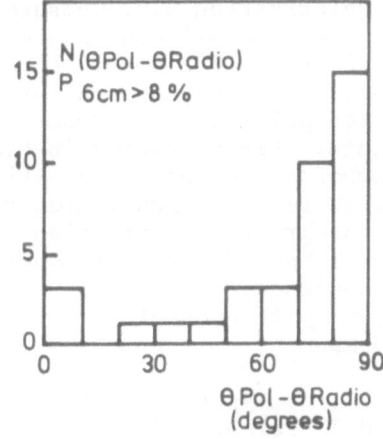

(Histogram includes only 3C sources with 6 cm polarization > 8%.
Where intrinsic polarization p.a. θ_{pol} is unknown, the measured
p.a. of the 6 cm polarization has been substituted. Typically,

Faraday rotation should not alter θ_{pol} by more than $\sim1\%$ at 6 cm
so the substitution is valid).
(iv)   Usually smooth but occasionally
quite blobby (e.g. 3C 61.1, 184.1, 234,
452, PKS 1004+13).
(v)   Usually roughly aligned but
occasionally one or more components
have completely different orientation (e.g. 3C 9, 69, 86A, 336,
390.3) (see Harris, 1974).
(vi)   Sometimes there is a bulge in
one component (e.g. 3C 285, 430).
(vii)   Sometimes two components have
same alignment but different from
orientation of source as a whole
(S-shape) (e.g. 3C 47, 4C 37.43).
(viii) In some cases hints of a similar
S-symmetry are also apparent between
the detailed features of opposite
components (e.g. Cygnus A, 3C 223).

d. Properties of outer hot spots:

(i)   Spectrum probably slightly flatter than for extended lobes
e.g. For Cyg A: $\alpha = -0.8$.   Corresponding electron lifetimes
$< 5 \times 10^4$ y.   At speed c electron takes $\sim 15 \times 10^4$ y. to travel
from nucleus to hot spot.   Therefore, processes within the hot
spots must be important to the particle acceleration mechanism.
(However, see Section 2 for many inherent assumptions.)
(ii)   Large fractional polarizations (e.g. $\sim 12\%$ for Cyg A).
(iii)   Polarization p.a. measurements suggest that magnetic field
in hot spots are aligned perpendicular to the source orientation.
This would explain a possible secondary maximum near $0°$ in the
integrated polarization p.a. histogram (section 3.2c (iii)).   The
integrated polarization of the whole source may consist of two
components.   First, a contribution from the extended lobes with
field *parallel* to the source axis.   Second, a part from the hot
spots with field *perpendicular* to the axis.   The p.a. indicated
by the integrated polarization would then depend on whether the
extended lobes or hot spots dominate.
(iv)   Hot spots are precisely aligned with compact nuclear component.

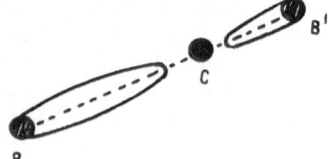

$B\hat{C}B' < 1°$ for Cyg A
        $< 0.5°$ for 3C 236

(v)   Strongest component usually nearer parent galaxy (Mackay,
1971).

e. Properties of compact nuclear components ('cores').

(i)   Usually flat spectrum ($-0.3 < \alpha < 0.3$). (Probably similar
to small variable sources such as 3C 345 and 3C 273B described
in Kellermann's lectures. Existence of radio cores was missed
until the last few years because of lack of high resolution
instruments at cm wavelength).
(ii)   Detected in most of closest sources. Probably present at
> 1% level in all strong radio sources but not yet observed due
to lack of sensitivity.
(iii)   Stronger sources tend to have stronger cores. (See Colla
et al. 1975.)
(iv)   Where structural information is available cores are extended along the source axis (e.g. Cyg A, 3C 111, 3C 236).   3C 236
is the only case where we have detailed information about the
structure of the core. See Fomalont and Miley (1975). Because
these are new results that place useful constraints on the mechanism
of double radio source production, I will present them in some
detail.
        3C 236 is the largest source (39' $\sim$ 6 Mpc) and contains the
largest core (0.8" $\sim$ 2 kpc). Core p.a. is 121° ± 4°. Source
p.a. is 122.5° ± 0.5°. In both cases westerly structure is
bigger. There is a fine structure (subcomponents) on a scale of
< 0.05" (100 pc).
        First, is the core the *relic* of a past event in the nucleus
which simultaneously produced the outer components of 3C 236 or
is it the product of *recent* nuclear activity? If it is a relic
it must have been confined for > 2×10^7 y, (i.e. $c\theta$). But minimum
internal pressure in subcomponents (< 100 pc) is $P_{min} \sim 10^{-7}$ dyne
cm^{-2}. For pressure balance:

Thermal confinement	$\rightarrow \rho > 10^{-22}$ gm cm^{-3}
Ram pressure confinement	$\rightarrow \rho > 10^{-21}$ gm cm^{-3}
Inertial confinement	$\rightarrow \rho > 10^{-25}$ gm cm^{-3}

These densities must be present over at least 2 kpc. In all cases
differential Faraday rotation should then wash out polarization.
However, source is observed to be $\sim$ 2% polarized at cm wavelengths.
Therefore, the core is the *product of recent nuclear activity*.
(See Section 2 for many assumptions of this argument.) Alignment
of 2 kpc core with 6 Mpc source implies that the orientation of
the nuclear powerhouse must have remained constant for at least
2×10^7 y. The directional properties are presumably associated
with the *rotational axis* of the nucleus or of a dense object within
it. (Not the rotational axis of galaxy as a whole since there is
no clear correlation between orientation of radio source and the
visible image of its parent galaxy - Gibson (1975).) Such aligned recurrent expulsions would be impossible in models such as the
'gravitational slingshot' in which expulsion takes place in the
equatorial plane.

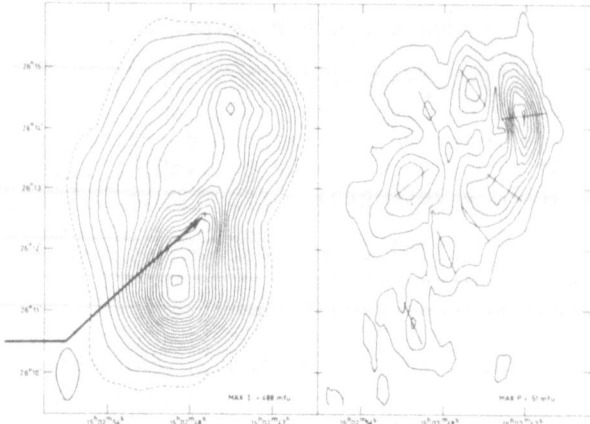

radio core
dominates at
higher frequen-
cies.

Fig. 7. 1.4 GHz total intensity and polarization contour map of
3C 310 (from Miley and van der Laan (1973)).

3.4 Relaxed symmetric doubles

About 10% of strong doubles sources are not highly collimated.
In the case of these blobby doubles there are no outer hot spots;
the brightness falls off gradually towards the edges. Relaxed
double sources are intrinsically more than an order of magnitude
less luminous but display many of the characteristics of collimated
symmetric doubles. Examples are Centaurus A, Fornax A, 3C 310
and probably Virgo A.

a. Main characteristics:

(i)   Two extended lobes, symmetrical about visible object.
Polarization measurements show complicated magnetic field structure
with some indications that the field directions are circumferential.
(ii)   Compact nuclear component (core) coincident with nucleus of
galaxy or QSO. Sometimes (e.g. for Centaurus A and Virgo A where
the sources are close enough to resolve their cores) the core
structure shows S-symmetry.

b. Possible explanations for relaxed doubles:

(i)   Dying double sources in which the energy supply to the components has been stopped causing the outer hot spots to be ex-
tinguished. However, the presence of compact nuclear components
in most relaxed doubles argues that the powerhouse of the source
is not dead.
(ii)   Sources in which the orientation of the nuclear powerhouse
has changed greatly during the source lifetime (e.g. due to precession). This could explain the S-symmetry in the structure of
the cores. We could then envisage the following sequence:

Axis of nuclear powerhouse constant →
large well collimated doubles (3C 236)

Slow precession of axis →
S-shaped symmetry between component
details (Cyg A)

Medium precession of axis →
2 components similarly aligned, different from source axis (3C 47)

Fast precession of axis →
S-symmetry in core. Hot spots cannot
form. Relaxed, double sources formed.

3.5 Head tail sources

a. Main characteristics:

(i)   High brightness radio head, C,
       coincident with visible galaxy.
(ii)   Low brightness radio tail, A,
       often stretching for hundreds
       of kpc.
(iii) Tend to occur in clusters of
       galaxies.

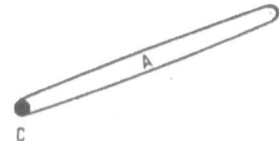

b. Most information from:

3C 129 which is located in an optically obscured region and has
the largest angular size (Fig. 8), and NGC 1265 in the Perseus
Cluster, the strongest source (Fig. 9).

Fig. 8. 1.4 GHz Westerbork contour map of 3C 129 from Miley (1973).

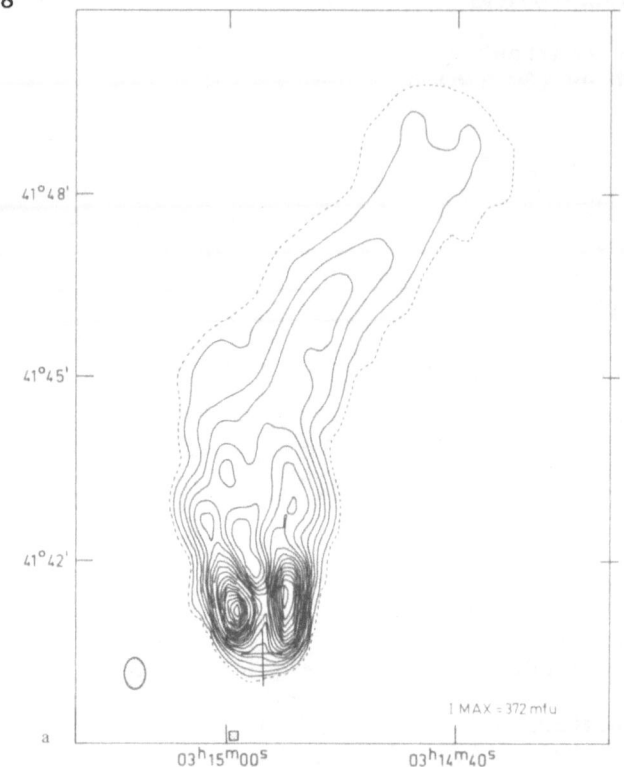

41°48'

41°45'

41°42'

I MAX = 372 mfu

a

03^h15^m00^s   03^h14^m40^s

Fig. 9.   1.4 GHz Westerbork contour map of NGC 1265 from Miley
(1973).   See Wellington et al. (1974), Miley et al. (1975).

c. Detailed properties (from NGC 1265 and 3C 129):

(i)   Head and tail are both double.
(ii)   Fractional polarization is larger in more relaxed extended
       tail regions of source (sometimes reaches $\sim$60%).
(iii) Spectrum is steeper for more relaxed extended tail regions
       of source.
(iv)   Compact nuclear component (core) almost always present.

d. Radio trail hypothesis (see Miley et al. 1972, Jaffe and Perola,
   1973).

First note that:

(i)   The above four properties are common to symmetrical double
radio galaxies and argue for a close relationship between double
and head tail sources.
(ii)   Head tail sources occur preferentially in clusters where the
medium might well be denser.
(iii) Three of the dozen or so head tail sources so far detected are
located in the Perseus Cluster whose galaxies have a dispersion

in radial velocities of $\sim$ 1600 km s^{-1}, larger than any other cluster known.

(iv) Galaxies travelling at these speeds would travel the length of a typical tail (few hundred kpc) within the radiative lifetime of the source deduced from synchrotron theory ($\sim$ 10^8 y).

The above considerations led to the suggestion that head tail sources are normal double radio galaxies distorted by *motion* of the parent galaxy through an intergalactic medium. The tail is then a *trail* defined by the orbit of the galaxy and motions of the intergalactic gas. In this case head tail galaxies are useful both for studying the evolution of sources and as a probe of the physical conditions in clusters.

e. More detailed properties – clues to source evolution:

(v)   Head of NGC 1265 (Wellington et al. 1974) shows pairs of discrete blobs strung out in semi circle which eventually merges

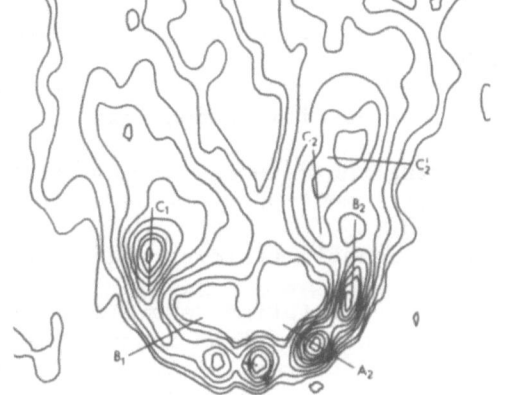

into tail. Suggests that nucleus undergoes recurrent activity giving birth to pair of blobs ($\sim$ 10^{55} erg) every few million years.

(vi) Central connection from back of semi circle to compact nuclear component. Possible trail of nucleus itself gently 'simmering' between large outbursts.

(vii) Tangential *magnetic field* in NGC 1265 is aligned *along tail* (Miley et al. 1975) This agrees with prediction of the magnetospheric model of radio trails (Jaffe and Perola, 1973) in which the galaxy is assumed to possess a dipole field which drags behind as the system moves through the medium (viz. earth's geomagnetic tail). Since in this picture the magnetic field lines on opposite sides of the tail go to opposite poles of the magnet, the line of sight magnetic field should change direction across the tail. There is evidence from the intrinsic rotation data that this is indeed the case.

(viii) In the heads of NGC 1265 and 3C 129 (Fig. 10) there are 'wiggles' of the same form in opposite arms. The wiggles which occur in both arms may be due to changes in the axis of the nuclear powerhouse (precession) as the galaxy moves through the medium. Alternatively they could be due to instabilities in the medium on a scale of tens of kpc.

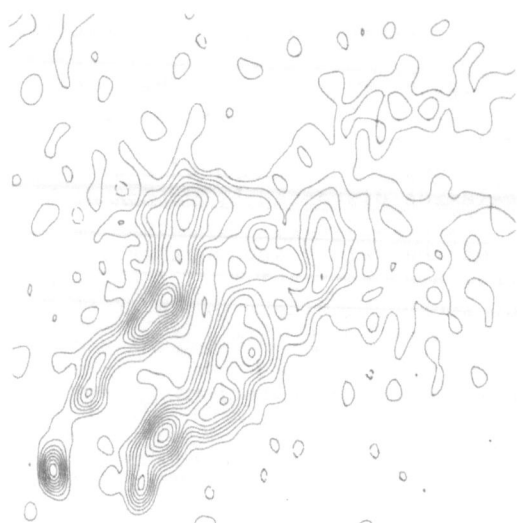

Fig. 10.   Unpublished Westerbork 5 GHz contour map of the head of
3C 129.

f. Physical conditions near radio trails:

(i)   Using the methods outlined in Section 2 (i.e. with numerous
assumptions) some physical parameters of head tail galaxies can
be derived.
Internal magnetic field strengths (equipartition) $H_{eq} \sim$ few micro-
gauss.
Corresponding lifetimes of relativistic electrons $t_r \sim 10^8$ y.
Internal densities (from rotation measures) $n_{int} \sim 10^{-4}$ to 10^{-3} cm^{-3}.
Total mass of thermal plasma inside source, $M \sim 10^{8.5}$ $M_\odot$.
(ii)   One can also use radio trails as a probe of the intra cluster
medium. Making the additional assumptions that the blobs are
retarded by ram pressure and that the tail is statically confined
by thermal pressure gives:
    External densities $\sim 10^{-3}$ cm^{-3}
    External temperature $\sim 10^8$K.
These values agree with those derived independently from a thermal
bremsstrahlung interpretation of the X-ray measurements.
(iii) Finally curvature on the scale of hundreds of kpc has been
observed in several head tail sources including NGC 1265 and 3C 129.
In most cases it cannot be simple gravitational bending (because
of the enormous masses required) and streaming motion within the
intergalactic medium must be involved.

3.6 Clues from morphologies of doubles and head tails

Let us summarize then in Table II the facts that have emerged from
our study of the structures of double and head tail sources. Some

of our deductions are more solidly grounded than others but as a whole we can draw a picture that is self consistent.

Table II

Evidence	Deduction
Widespread presence of radio cores. Double symmetry. Presence of hot spots. Alignment of hot spots with core.	Radio source is powered from nucleus. Energy is transmitted from nucleus in two exactly opposite directions, collimated to a few degrees.
Discreteness of NGC 1265 head.	Activity in the nuclear powerhouse can be intermittent.
Alignment of 3C 236 core with outer double.	In some cases the axis of the nuclear powerhouse is constant to a few degrees over the lifetime of the source ($\sim 10^8$ y).
S-shaped symmetry.	Sometimes the axis revolves during the source lifetime producing the mirroring effects observed. Fast revolution could produce relaxed doubles with S-shaped cores.
Confinement of sources. Existence of head tail.	Sources are imbedded in intergalactic medium. Sometimes motion of galaxy through dense intra cluster medium distorts normal double sources into radio trails.
Magnetic field of NGC 1265.	Galaxy has magnetosphere which is dragged along the trail.

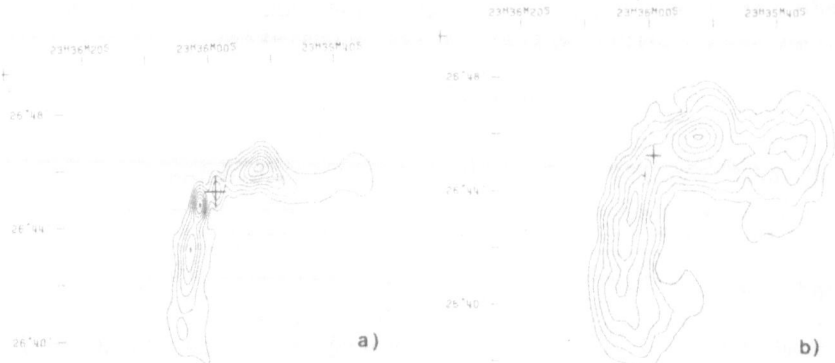

Fig. 11. a) 1.4 GHz Westerbork contour map of 3C 465 taken from
Miley and van der Laan (1973), but showing fewer contours. Source
is identified with the double galaxy NGC 7720 (cross) in Abell
Cluster A 2634. b) 0.6 GHz unpublished Westerbork measurements.

3.7 Complex sources

About 10% of the observed structures do not fall into any of the
morphological classes discussed in Table II. However, if we take
into account the effects of projection and superposition most of
these complex beasts can be explained within the general scheme
outlined in Table II.
     Take a few examples:
(i)   3C 465 - Probably intermediate case between double and
head tail source. There is significant transverse motion of galaxy
through medium but this motion is relatively less important than
in the case of a head tail. Therefore, arms of radio tail spread
out. Further evidence that this is the case comes from unpublished
0.6 GHz Westerbork measurements (Fig. 11b) which show emission from
between the arms, i.e. from the line defined by the assumed trajectory of the galaxy.

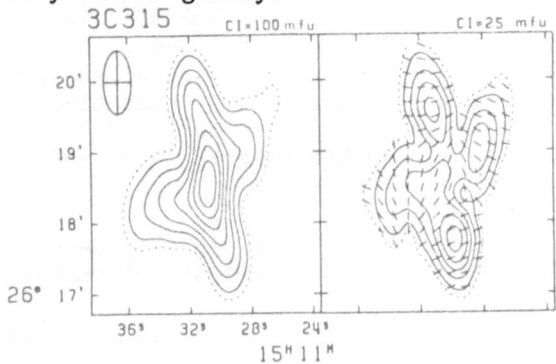

Fig. 12. Westerbork 1.4 GHz map of 3C 315 from Högbom and Carlsson (1974).

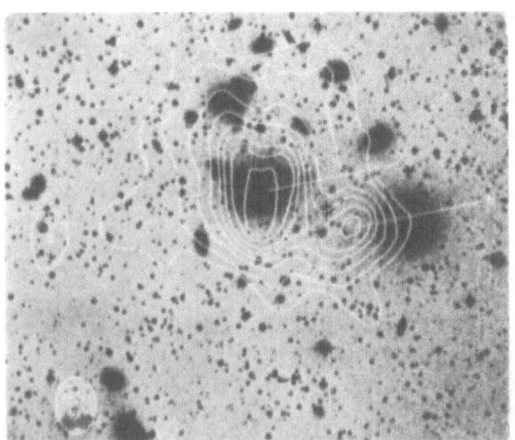

Fig. 13.   1.4 GHz Westerbork map of the halo of NGC 1275 at the centre of the Perseus Cluster. The radio core which coincides with the
optical nucleus has been subtracted, from Miley and Perola (1975).

(ii) Cross sources (Fig. 12) - At least two 3C sources, 3C 315 and
and 3C 386 have structures with cross shaped features. We can explain this if the nuclear powerhouse changes its orientation and
undergoes recurrent activity during the source lifetime. If the axis
precesses and undergoes two main outbursts a cross-shaped source
would result.
(iii) NGC 1275 (3C 84) (Fig. 13) - The amorphous form of the halo may
mirror the distribution of thermal plasma at the centre of this rich
cluster.

3.8 Selection effects and morphology

Instrumental limitations govern the quality of the observed maps and
mean that in some cases these maps may only show the "tips of the
iceberg". Three of the most important of these selection effects are:
(i) Maps of strong sources are usually dynamic range limited. Then
one misses structure weaker than a certain *fraction* of the strongest
intensity in the field. The fraction depends on the distance from
the strongest source.
(ii) Maps of weak sources will be noise limited. In this case components weaker than a given *absolute* intensity will be missed.
(iii) Spectral variations across the source may mean that the observed morphology depends on the observing frequency and/or the frequency
at which the radiation was emitted.
Instrumental limitations such as these,introduce several unwanted
selection effects into any analysis of source structures.

a. Misclassification of morphology

Take some examples:

(i)   Symmetric double source with very unequal outer components
could easily be misclassified as a head tail (if it had a strong
radio core) or unresolved.
(ii)   A few quasars such as 3C 273 appear to consist of a strong
radio core and only one other component (D2 quasars). It is not
yet known whether these constitute another class of (assymmetric
double) structure or whether a second outer component exists and
has simply been missed.

b. Influence on optical identifications

Traditionally a source was usually regarded as optically identified
if its radio centroid coincided with a visible object. For extended sources this is an unsatisfactory criterion. It discriminates
against highly assymmetric sources such as head tails. Nowadays
another useful method by which an extended source can be identified
is via the detection of a small radio core within it. The position
of this core can be determined to better than 1" and the search
for an optical identification is narrowed to this region. Clearly
such identifications will be missed if the core is undetected (e.g.
for reasons of dynamic range or sensitivity).

c. Source statistics

The above instrumental limitations make it extremely difficult to
obtain a homogeneous sample of morphologies. Beware of doing complicated statistics based on the assumptions that your data are
homogeneous.

3.9 The Future

Despite the danger of crystal ball gazing one can hazard a few
safe guesses for the next decade.
(i)   The large scale intensity and polarization structures of
many more sources will be mapped. Our sample will hopefully be
augmented by a new array in the Southern hemisphere. Some statistics of magnetic field distributions in the extended components
will become available. Also there will be information about
how detailed source morphology (including magnetic field distributions) changes as a function of redshift.
(ii)   Some may dare to trend the hitherto untouched field of radio
structure measurements at < 100 MHz.
(iii) The hot spots in many sources will be mapped by the VLA.
(iv)   Increasing use of VLBI should relate the detailed structure
of the radio cores to that of the extended components. Cross
fertilization of these radio data with observations from other regions of the e.m. spectrum (e.g. next generation of X-ray satellites)
may bring us a step nearer to solving the fundamental problem of
why extragalactic sources are double and how the tremendous energies
are generated.

REFERENCES

Branson, N.J.B.A., Elsmore, B., Pooley, G.G., and Ryle, M., 1972, M.N.R.A.S., 156, 377.
Christiansen, N.N., 1953, Nature, 171, 831.
Cohen, M.H., Gundermann, E.J., and Harris, D.E., 1967, Astrophys. J. 150, 767.
Colla, G., Fanti, C., Fanti, R., Gioia, I., Lari, C., Lequeux, J., Lucas, R., Ulrich, M.-H., 1975, Astron, and Astrophys., 38, 209.
Cooper, B.F.C., Price, R.M., and Cole, D.J., 1965, Aust. J. Phys., 18, 589.
Ekers, R., 1969, Aust. J. Phys. Suppl. 6.
Ekers, R., 1975, Structure and Evolution of Galaxies, G. Setti, editor, D. Reidel Publishing Co., Dordrecht, Holland, pp. 217-246.
Elsmore, B. and Mackay, C.D., 1969, M.N.R.A.S., 146, 361.
Fanaroff, B.L. and Riley, J.M., 1974, M.N.R.A.S. 167, 31p.
Fomalont, E.B., 1971, Astron. J., 76, 513.
Fomalont, E.B. and Miley, G.K., 1975, Nature, 257, 99.
Fomalont, E.B. and Wright, M.C.H., 1974, Galactic and Extra-Galactic Radio Astronomy, Chap. 10, Ed. Verschuur and Kellermann, Springer-Verlag.
Gardner, F.F. and Whiteoak, J.B., 1963, Nature, 197, 1162.
Getmansev, G.C. and Ginzburg, V.L., 1950 Zh. Eskperim. Theor. Fiz. 20, 347.
Gibson, D.M., 1975, Astron. and Astrophys. 39, 377.
Graham, I., 1970, M.N.R.A.S., 149, 319.
Hargrave, P.J. and Ryle, M., 1974, M.N.R.A.S., 166, 305.
Harris, A., 1972, M.N.R.A.S., 158, 1.
Hazard, C., Mackey, M.B. and Shimmins, A.J., 1963, Nature, 197, 1037.
Hewish, A., Scott, P.F. and Wills, D., 1964, Nature, 203, 1214.
Hewish, A., Readhead, A.C.S. and Duffett-Smith, P.J., 1974, Nature, 252, 657.
Hoerner, S. von, 1964, Astrophys. J., 140, 65.
Högbom, J.A., 1974, Astron. and Astrophys. Suppl., 15, 417.
Högbom, J.A. and Carlsson, I., 1974, Astron. and Astrophys., 34, 341.
Hogg, D.E., 1969, Astrophys. J., 155, 1099.
Hogg, D.E., Macdonald, G.H., Conway, R.G. and Wade, C.M., 1969, Astron. J., 74, 1206.
Jaffe, W.J. and Perola, G.C., 1969, Astron. and Astrophys. 26, 423.
Kapahi, V.K., 1975, M.N.R.A.S., 172, 513.
Kapahi, V.K., Joshi, M.N. and Sarma, N.V.G., 1974, Astron. J. 79, 521.
Little, L.T. and Hewish, A., 1968, M.N.R.A.S., 138, 393.
Macdonald, G.H., Kenderdine, S. and Neville, A.C. 1968, M.N.R.A. S., 138, 259.
Macdonald, G.H. and Miley, G.K., 1972, Astrophys. J., 164, 237.

Mackay, C.D., 1969, M.N.R.A.S., 145, 31.
Mackay, C.D., 1971, M.N.R.A.S., 154, 209.
McVittie, G.C., 1965a, General Relativity and Cosmology, Chapman and Hall.
McVittie, G.C., 1965b, Astrophys. J., 142, 1637.
Mills, B.J. and Little, A.G., 1953, Aust. J. Phys. 6, 272.
Mitton, S., 1972, M.N.R.A.S. 155, 373.
Mitton, S. and Ryle, M., 1969, M.N.R.A.S., 146, 221.
Miley, G.K., 1971, M.N.R.A.S., 152, 477.
Miley, G.K., 1973, Astron. and Astrophys., 26, 413.
Miley, G.K. and Laan, H. van der, 1973, Astron. and Astrophys., 28, 359.
Miley, G.K. and Perola, G.C., 1975, Astron. and Astrophys., in press.
Miley, G.K., Perola, G.C., van der Kruit, P.C. and van der Laan, H., 1972, Nature 237, 269.
Miley, G.K., Wellington, K.J. and Laan, H. van der, 1975, Astron. and Astrophys., 38, 381.
Mills, B.J. and Little, A.G., 1953, Aust. J. Phys. 6, 272.
Mitton, S., 1972, M.N.R.A.S., 155, 373.
Mitton, S. and Ryle, M., 1969, M.N.R.A.S., 146, 221.
Pacholczyk, A.G., 1970, Radio Astrophysics, Chap. 7, W.H. Freeman and Co.
Pooley, G.G. and Henbest, S.N., 1974, M.N.R.A.S. 169, 477.
Refsdal, S., Stabell, R. and de Lange, F.G., 1967, Mem.R.A.S., 71, 143.
Riley, J.M., 1972, M.N.R.A.S., 157, 349.
Riley, J.M. and Branson, N.B.J.A., 1973, M.N.R.A.S., 164, 271.
Roeder, R.C., 1974, Nature, 255, 124.
Ryle, M., 1952, Proc.Roy.Soc., A-211, 351.
Ryle, M. and Windram, M.D., 1968, M.N.R.A.S., 138, 1.
Scheuer, P.A.G., 1962, Aust. J. Phys., 15, 333.
Schlizzi, R.T. and McAdam, W.B., 1975, Mem.R.A.S. 79, 1.
Swarup, G., 1975, M.N.R.A.S., 172, 501.
Turland, B.D., 1975, M.N.R.A.S., 170, 281.
Wardle, J.F.C. and Miley, G.K., 1974, Astron. and Astrophys. 30, 305.
Wellington, K.J., Miley, G.K. and van der Laan, H., 1973 Nature, 244, 502.
Willis, A.G., Strom, R.G. and Wilson, A.S., 1974, Nature, 250, 625.

COMPACT RADIO SOURCES IN QUASARS AND GALACTIC NUCLEI

Kenneth I. Kellermann

National Radio Astronomy Observatory*
Green Bank, West Virginia

1. INTRODUCTION

Compact radio sources are associated with quasars, and with the nuclei of galaxies, particularly active galaxies such as N and Seyfert type, but including elliptical and spiral galaxies as well. Their radio luminosity ranges from $\sim 10^{39-40}$ ergs/sec in nearby galaxies to $\sim 10^{46}$ ergs/sec in the large red shift quasars (assuming of course that their red shifts are cosmological). Compact sources are not rare or unusual, but because they become self absorbed at longer wavelengths, they are very weak below the cutoff frequency, and do not appear in most radio catalogues which are based on long wavelength surveys. More than half of the sources found in short wavelength surveys are compact. Also a number of extended sources contain weak compact sources at their center.

2. COMPACT SOURCE SPECTRA

A uniformly bright source of synchrotron radiation from ultra relativistic electrons becomes self absorbed at a frequency ($\tau \sim 1$)

$$\nu_m \sim 10 \; S_{Jy}^{2/5} \; \theta_{m \; sec}^{-4/5} \; B_{Gauss}^{1/5} \; (1 + z) \quad GHz \qquad (2.1)$$

e.g. Kellermann (1974). For $\nu \gtrsim 2 \nu_m$, the source is transparent

---

* Operated by Associated Universities, Inc., under contract with the National Science Foundation.

G. Setti (ed.), The Physics of Non-Thermal Radio Sources, 27–40. All Rights Reserved.
Copyright © 1976 by D. Reidel Publishing Company, Dordrecht-Holland.

and the flux density $S \propto \nu^{\alpha}$ where $\alpha = (1 - \gamma)/2$ and γ is the index of the relativistic electron energy distribution $N(E) \propto E^{-\gamma}$. Typically $\alpha \sim -0.8$ and $\gamma \sim 2.6$. For $\nu \gtrsim \nu_m/2$, the source is opaque and $\alpha = 2.5$. In general, the observed values of α for compact sources lie in the range -0.4 to $+1$. Values as steep as 1.5 or 2 are observed infrequently. The apparent absence of the theoretical value of 2.5 has been used as an argument against interpreting the compact source spectra as a result of synchrotron self absorption. However, this is easily understood, if the source is not uniformly bright, and different parts of the source become opaque at different wavelengths. Thus instead of the theoretically sharp cutoff, there is a gradual cutoff in the spectrum which asymptotically approaches a spectral index of 2.5. But this value is not reached until the source is entirely opaque, which is difficult to observe since the flux density is then too small. The observed departure of HII regions from the theoretical blackbody index of $+2.0$ is explained in a similar manner. Models of inhomogeneous radio spectra have been discussed by Condon and Dressel (1974).

The measured angular dimensions of compact sources require self absorption to occur unless the magnetic fields are un-realistically low. Observed values of the spectral index $\alpha \gtrsim 0.3$ is always a sign of a compact source. In the compact sources, generally α is determined by the distribution of opac-ity, and not by a low value of γ. $1 < \gamma < 1.5$ $(-0.25 < \alpha < 0)$ is observed in the transparent part of compact source spectra (at short wavelenths), but never in the extended sources.

The smallest radio components which are observed have $\theta \lesssim 0.25$ milli arc sec. These have spectral peaks at $\nu_m \gtrsim 10$ GHz. Even smaller sources may exist, but these would be self absorbed at higher frequencies and not visible at cm wavelengths unless they have an extended companion. Surveys at $\lambda < 2$ cm $(\nu > 15$ GHz$)$ are required to find sources with $\theta < 0.2$ milli arc sec, but currently available sensitivities make such a survey difficult.

The smallest angular size which may be observed at any wave-length is set by inverse Compton cooling which limits the peak brightness temperature $T_B \lesssim 10^{11-12}$ °K. In sources with greater brightness temperature, the relativistic particles would rapidly lose their energy due to inverse Compton radiation at IR, opti-cal, or X-ray emission, rather than synchrotron emission at radio wavelengths.

3. STRUCTURE OF THE COMPACT SOURCES

In recent years, it has become increasingly clear that a large source of energy exists in the nuclei of galaxies and in quasars which frequently gives rise to an intense source of radio, IR, optical, or X-ray emission. Observations with extremely great

angular resolution are required to study these phenomena, which
have a linear extent less than 100 pc, and sometimes as small as
one light year or less.

Such high resolution observations are possible only at
radio wavelengths where the fluctuations in path length through
the earth's atmosphere are small compared with a wavelength, and
where coherence can be maintained over vast distances. Thus at
radio wavelengths it is possible to build very large diffraction
limited instruments, with angular resolutions better than 1 milli
arc second. Until a few years ago, this important advantage of
radio astronomy was not generally appreciated. Indeed, it was
widely believed that due to the relatively long wavelength involved the resolution of radio telescopes was fundamentally
orders of magnitude worse than for optical telescopes. In fact,
the reverse is true, and long baseline radio interferometers are
being used to study quasars, galactic nuclei, interstellar molecular masers, pulsars, and radio stars with unprecedented angular
resolution.

3.1 Long baseline interferometry

Before discussing the high resolution observations of compact
radio sources, we summarize briefly the principles of radio
interferometry, in particular the use of independent oscillator
tape recording interferometers to extend the baselines to
thousands of miles. More complete descriptions are given by
Fomalont and Wright (1974) and by Cohen (1969).

Any two-element interferometer samples a single point in
the two dimensional Fourier transform of the angular brightness
distribution. As the earth rotates, the interferometer sweeps
out an ellipse in the (u,v) or Fourier transform plane. By
using multiple element systems, a series of ellipses covers the
Fourier transform plane, and a formal Fourier transform may be
used to reconstruct an image of the radio source. Conventional
radio arrays with dimensions of the order of a few km are used
in this way to obtain radio source images with resolutions of
the order of a few arc seconds.

The very long baseline interferometer (VLBI) systems are
similar, in principle, except that the cables are replaced by
tape recordings at each end of the interferometer which are
later correlated to produce the interference fringes. In practice, however, there are several factors which limit the perform-
ance of VLBI systems. These are:

1) The antennas are usually not located at the optimum
locations to uniformly sample the (u,v) plane.

2) There are generally too few antennas to adequately
cover the (u,v) plane.

3) It is difficult to obtain the necessary phase information so that a formal Fourier transform is not possible. The
phase information is lost due to:

   a)   instabilities in the independent oscillators used as
the phase reference at each end of the baseline;
   b)   fluctuations in the path length through the atmosphere above each antenna;
   c)   uncertainties or changes in the interferometer baseline due to:

   i)   inadequate geodetic data over transcontinental
         and intercontinental baselines;
   ii)   solid earth tides;
   iii) changes in the direction (polar motion) or rate
         of rotation (UT1) in the earth's axis;
   iv)   plate motion or continental drift.

   In principle, given reasonably stable local oscillators,
and sufficient data, it is possible to solve for each of the
above effects, as well as the source visibility phase. Errors
in the relative visibility phase may be reduced by using a nearby reference source.

   Until the present, however, most VLBI data has been interpreted by model fitting. In general, one looks for the simplest
geometric model which fits the observed data. The models derived
in this way are not unique, but earlier experience with phaseless
data using conventional connected interferometers suggests that
the models derived in this way will give a good approximation to
the true brightness distribution.

   Most of the earlier VLBI observations were made using a
single interferometer baseline. This was sufficient only to
determine the over-all source dimensions. More recently, the
observations have used 3 to 5 stations giving up to 10 independent interferometer baselines with reasonably uniform coverage of
the (u,v) plane. Observations are currently being made at wavelengths between 1 cm and 1 meter, with the emphasis on the
shorter wavelengths where the resolution is greatest and where
there is minimum distortion from the ionosphere or interstellar
medium.

3.2 Properties of compact sources

The primary results of very long baseline interferometer observations of compact radio sources may be summarized as follows:
   1)   The compact sources show a wide range of angular and
linear size which essentially form a continuous sequence with
the extended sources.
   2)   Many sources contain several compact components. The
smallest ones are strongest at shorter wavelengths. This is
expected as a result of self absorption. The smallest components have measured angular dimensions $\theta \lesssim 0.25$ milli arc sec in
angular size, or about 1 light month in the closer galaxies.
   3)   There are no completely unresolved sources at the
highest angular resolution currently available. But there are

unresolved components of sources with dimensions $\theta \lesssim 0.25$ milli
arc seconds. There may be much smaller sources in the sky, but
these are likely to be opaque for $\lambda > 1$ cm, and so will only be
detected in sky surveys at $\lambda \lesssim 1$ cm.

4) The typical peak brightness temperature is about 10^{11} °K
-- close to the value expected from synchrotron emission and inverse Compton cooling. There is no evidence for brightness
temperatures in excess of 10^{12} °K, as would be expected from a
coherent emission process. However, it would be difficult to
measure such high values from earth based interferometers.

5) The magnetic field strengths calculated from the angular size and self absorption cutoff frequency is typically in
the range 10^{-6} to 10^{-4} Gauss.

6) When examined with sufficient angular resolution, the
structure of the compact sources is complex and generally consists of two or more spatially separated components. In fact,
the structure of the compact sources is remarkably similar to
that of the extended sources except for a scale factor of the
order of 10^5. This is particularly remarkable in view of the
very different conditions affecting the evolution of the two
types of sources:

a) The total energy contained in the compact sources
in the form of relativistic particles and magnetic fields is
$\lesssim 10^{55}$ ergs compared with $\sim 10^{60}$ ergs in the extended sources.

b) The density of the medium surrounding the compact
sources found in galactic nuclei or quasars is very much greater
than intergalactic densities associated with the extended
sources.

7) Compact sources are never found within the extended
radio lobes; but, only at the nucleus of the identified galaxy
or quasar. This does not support models where relativistic
particles are accelerated in situ, but rather is consistent with
acceleration in the galactic nucleus or quasar.

8) There appears to be no difference between the compact
sources found in radio galaxies or quasars.

9) In those sources where there is both an elongated or
double extended source and a central elongated compact source,
the directions of elongation are nearly aligned suggesting a
common mechanism which is effective over a range from a few pcs
to a few hundred kpc in determining the direction of elongation.

These properties are illustrated below with reference to
particular sources. More detailed discussions are given in
recent papers by Cohen, et al. (1975), Kellermann, et al. (1976),
Schilizzi, et al. (1975), Shaffer, et al. (1975), Pauliny-Toth,
et al. (1975), and Wittels, et al. (1975).

3.3 Examples of compact source structure

1) <u>1633+38</u> (quasar) is one of the smallest sources known

($\theta \sim 0.4$ milli arc seconds). This source also has a very high
self absorption cutoff frequency, $\nu_m \sim 12$ GHz consistent with its
small size. The magnetic field, $B \sim 3 \times 10^{-3}$ Gauss, is rather
large as in other sources of very high surface brightness.

2) OJ 287. Most of the flux at short cm wavelengths is
contained in a very small component, $\theta \sim 0.4$ milli arc seconds.
There is also a somewhat larger halo of ~ 2 milli arc seconds
which contains an increasingly greater fraction of the total flux
density at longer wavelengths. The small core is highly variable, while the flux of the halo remains relatively constant.

3) BL Lac has been observed since 1971. This source is
extremely variable with outbursts every few months. The source
has remained elongated near pa ~ 0 during the past 4 years although the lifetime of an individual outburst is considerably
less than one year. The highest resolution observations show a
double source structure consisting of a halo and a smaller core
separated by ~ 1.25 milli arc seconds oriented near pa ~ 0. The
different outbursts apparently are somewhat separated in space,
but occur along a preferential axis. It is not clear what deter-
mines the direction of this axis, but it is unlikely to be the
magnetic field, since polarization measurements indicate that
this rotates relatively rapidly.

4) M87. This is one of the closer radio galaxies. Its
nucleus contains a complex radio sources of ~ 0.5 arc seconds
in extent. The core of this nucleus is only 0.4 milli arc seconds ($1\frac{1}{2}$ light months) in extent. Quite surprisingly, this
very small component has not varied significantly.

5) NGC 5128 is the closest radio galaxy (Centaurus A). The
radio emission extends over a region nearly one Mpc across. The
galaxy contains the well known dust lane and lies between two
radio components separated by about 50 kpc. At the nucleus of
the galaxy is a very compact radio source less than 1 milli
arc seconds (1 light month) across. Because NGC 5128 is located
so far south, detailed interferometric data is not available.
However, the very high spectral cutoff frequency (~ 30 GHz) and
rapid observed time variations, suggest a size ~ 0.25 milli arc
seconds (1 light week). A greater size implies a large magnetic
field and very short radiation lifetime. A smaller size would
mean excessive inverse Compton scattering and again a very short
lifetime. The nucleus of NGC 5128 also contains a bright infra-
red source, one of the strongest extragalactic X-ray sources,
and the only detected γ-ray source. It is clearly a center of
violent activity.

6) Cygnus A is one of the brightest extended double radio
sources with an absolute luminosity $\sim 10^4$ times greater than
that of NGC 5128. Like NGC 5128 there is a small compact radio
source coincident with the identified galaxy. This compact
source is elongated along the same direction as the line joining
the extended components. The dimensions of the radio core is
~ 2 by 1 pc compared with 200 kpc for the separation of the

extended components. Similar alignments of compact components
with the extended double structure are found in a number of
other radio galaxies and quasars including 3C 111, 3C 147 and
3C 154.

For some years it has been considered that Cyg A was
identified with a pair of interacting galaxies. However, it is
unlikely that we would find a compact source in the region between two galaxies, and by analogy with NGC 5128, it is likely
that the apparent "pair of galaxies" is, in reality, a single
galaxy, covered by a dust lane, with a compact radio source at
the nucleus.

7) NGC 1275 (3C 84) contains a unique compact radio source
which is both sufficiently large at short centimeter wavelengths
that it can be resolved into many picture elements, and at the
same time sufficiently strong that there is adequate sensitivity
to make accurate measurements even when the source is highly
resolved.

At centimeter wavelengths, the flux density is dominated by
a component (A) which is about 6 milli arc seconds (3 pc) across
and consists of three apparently stationary centers of emission
aligned along position angle ∿ -9 degrees. Another component (B)
which dominates the flux at ∿ 30 cm is ∿ 0.06 (30 pc) arc seconds
across, and is also aligned in nearly the same direction. A much

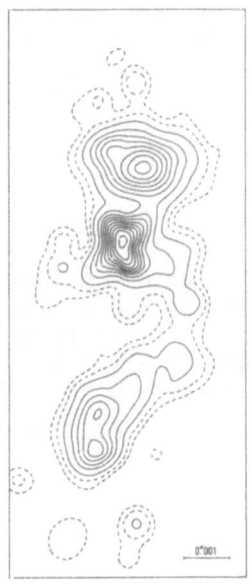

Fig. 1. NGC 1275: brightness contours taken from Pauliny-Toth
et al. (1975). Component A.

larger component (C) is $\sim$ 5' arc in extent, and the optical
filaments are also extended in the same direction.

This remarkable alignment of components with dimensions
ranging from a few pc to a few hundred kiloparsecs suggests a
common mechanism for the orientation of the components which is
effective over a wide range of distances and time as well.

The 200 kpc component must be at least 10^6 years old and
more likely 10 or 100 times this. The small 3 pcs triple source
on the other hand is only $\sim$ 15 years old since observations by
Heeschen in 1960 showed this source to be much weaker at that
time.

High resolution data taken since 1972 show that each of the
3 major centers of Component A has remained essentially stationary over the past 3 years while at the same time their relative
flux density and possibly their size has varied significantly.

4. VARIABLE RADIO SOURCES

Essentially all compact sources which have been observed for
several years show flux density variations on time scales ranging
from a few months to a few years. In general, at those frequencies where the source is opaque ($\tau > 1$) the flux density increases with time; in the transparent region ($\tau < 1$) it decreases
with time. This is the basis of the expanding source model which
assumes that the source starts out as a small cloud of relativistic particles which expands. Upon expansion the magnetic field
and the particle energy decreases, producing an increase in flux
where $\tau \gg 1$ and a decrease where $\tau \ll 1$. To be more quantitative, it is necessary to make some assumptions about the kinematics, the manner in which B varies, the acceleration of particles, and their rate of energy loss due to synchrotron emission
and to inverse Compton scattering.

In the simplest case we assume that:

    a) the relativistic particles are produced instantaneously
in an infinitesimal volume of space;
    b) the particle spectrum is of the form $N(E) = KE^{-\gamma}$;
    c) the expansion is linear ($\theta \propto t$);
    d) the magnetic flux is conserved ($B \propto \theta^{-2} \propto t^{-2}$).

Then it may easily be shown that (e.g. Kellermann and Pauliny-
Toth (1968)):

    1) if $\tau \gg 1$, $S \propto t^3$
    2) if $\tau \ll 1$, $S \propto t^{-2\gamma}$
    3) $S_{peak} \propto \nu^{(7\gamma+3)/(4\gamma+6)}$
    4) $\nu_{peak} \propto t^{(4\gamma+6)/(\gamma+4)}$

All of the above assumptions can be challenged, however.
For example:

a) The particle acceleration must last a finite time and the initial source dimensions must be finite.

b) Even if the energy distribution is initially a power law, energy losses will distort the simple distribution.

c) The rate of expansion is likely to be decelerated, particularly if it occurs in a galactic nucleus or within a quasar.

d) The expansion may occur in a fixed magnetic field.

However, the simple model provides a basis for quantitative analysis. Departures from the predictions of the simple model are not to be thought of as evidence against the general class of expanding source models, but rather as information about the parameters of the model. This is especially true if the rate of change of size as well as flux density is measured.

In practice, it is difficult to quantitatively analyze the data, because the individual events are superimposed both in frequency and in time, and it is impossible to determine accurately the parameters of an individual burst. With the limited data available, however, the variations conform at least qualitatively to the expanding source model; in a few cases where there has been one or more relatively large isolated outbursts, there is good quantitative agreement. Recent data on radio source variations is given by Medd et al. (1972), Dent and Kojoian (1972), and Dent and Hobbs (1973).

In the transparent part of the spectrum of compact sources $\alpha \sim -0.25$ ($\gamma = 1.5$). It is likely that this is the index of the "production spectrum" and that the value of ~ -0.8 normally found in extended sources is the result of subsequent energy losses.

The major quantitative discrepancies between the simple expanding source model and the observed data are

a) At wavelengths $\lambda \lesssim 1$ cm, the amplitude of the outbursts are nearly independent of wavelength, and there is no observed time delay at longer wavelengths. This has been used as an argument against the expanding source model. Rather it may be interpreted as evidence for a finite initial size. For $\lambda \lesssim 1$ cm, most sources are apparently always transparent, and the flux density increases with time due to the increase in the number of relativistic particles, rather than as a result of the change in size.

b) In some sources, (e.g. 3C 454.3 and BL Lac) the variations observed at decimeter wavelengths are considerably greater than predicted on the basis of the simple expanding source model. Again, this may be interpreted as being due to a separate larger variable component, which is transparent at centimeter wavelengths.

The real problem in accepting the universal applicability of the expanding source model arises when the apparently great distance of the variable quasars is considered.

Consider a source where the flux density changes significantly on a time scale τ. Assuming that the dimensions do not
exceed the length traveled by light during a time τ, and that
the distance of the quasar is given by the measured red shift,
then an upper limit to the angular size may be calculated. The
corresponding lower limit to the peak brightness temperature
sometimes exceeds by a considerable amount the upper limit of
10^{12} degrees set by inverse Compton scattering. This problem is
particularly acute in those sources where large rapid variations
have been observed at relatively long decimeter wavelengths, and
has led to the suggestion that at least in some sources the
radio emission is not due to ordinary synchrotron emission, but
is the result of some coherent process; or else that these
quasars are much closer than the distance corresponding to their
red shift. The magnitude of the discrepancy is, however, very
sensitive to the specific source geometry and the quantitative
definition of the characteristic time, τ, and it has not been
convincingly established that the observations cannot be explained within the framework of conventional synchrotron theory
and the conventional cosmological interpretation of the quasar
red shifts.

An elegant way out of this dilemma was suggested some years
ago by Rees (1967). He pointed out that if the source is expanding with a velocity close to the velocity of light, the differential travel time of signals originating in different parts of
the source will give the illusion of expansion with a velocity
greater than $\underline{c}$ and so the angular size is greater and the brightness temperature less than estimated from the simple light travel
argument given above.

Because of the great interest in the kinematics of variable
radio sources, there has been considerable effort to measure
directly the change in angular structure using VLBI techniques.
Most of the well resolved sources have a complex brightness
distribution and contain multiple components, and the observed
changes in structure is likewise complex. Individual components
may move, expand, or vary in flux.

Because the VLBI models are based on limited (u,v) coverage
and no phase data, the results are not unique. In general, the
simplest geometric model is chosen which fits the data; but this
does not always lead to the simplest physical model.

Some examples of variable radio structure are listed below.
See the papers by Schilizzi et al. (1975), Shaffer et al. (1975),
and Pauliny-Toth et al. (1975) for more details.

a) 2134+00 (quasar). This is a three component source
whose flux density has remained relatively constant for several
years. None of three components appears to have changed its
flux density, size, or relative position significantly.

b) 3C 454.3 (quasar). This source is quite variable. It

consists of a halo, plus an unresolved ($\theta \lesssim 0.25$ milli arc seconds) variable core which becomes relatively more prominent at shorter wavelengths.

c) 4C39.25 (quasar). This is a variable double source with a separation about 2 milli arc seconds. The flux density of both components has been increasing with time during the past few years, but one of them is changing more rapidly than the other. The apparent separation has not changed significantly.

d) 3C 120 (galaxy). This is a complex source which has had repeated large intensity bursts. Each outburst appears to produce two components which separate with an apparent velocity close to the velocity of light (Seielstad 1974).

e) 3C 84 (galaxy). This complex source which has been discussed above, has not changed its overall size, nor does there appear to be any significant component motion. The intensity and perhaps the size of one or more of the three major components has, however, varied during the past few years.

f) 3C 273 (quasar). Observations of the quasars 3C 273 and 3C 279 made in 1970 and 1971 were interpreted as indicating a double component source which was separating at an apparent velocity of at least several times the speed of light. It was realized, at the time, that the limited data then available could also be interpreted as a more complex geometric model having stationary components which varied in flux.

More detailed observations made in the spring of 1972 clearly showed a triple component configuration with a separation of 2.2 milli arc seconds. Reinspection of the older more limited data showed that all of the then available data could be interpreted in terms of a fixed component model having three components with variable intensity. Later observations, however, give

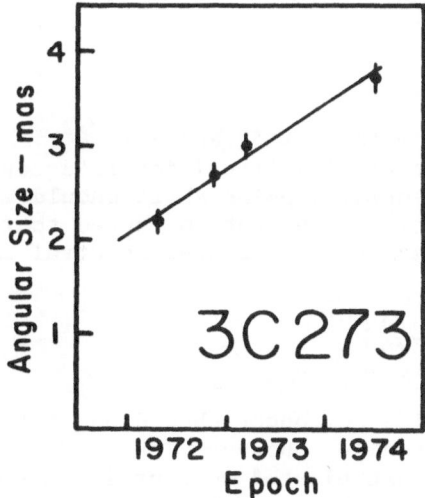

Fig. 2. 3C 273: Rate of expansion.

convincing evidence that the three components are separating
with an angular velocity of 0.8 milli arc seconds per year or
$\sim$ 5 c (H = 50) for the linear velocity of each of the outer components with respect to the center. At least one of the individual components itself appears to also be expanding with a
comparable velocity.

Three categories of explanation have been suggested to explain what must be considered convincing evidence for component
motion.

a) The source is much closer than indicated by the
cosmological interpretation of the red shift and the currently
accepted value of the Hubble constant.

b) "Faster-than-light" motion really occurs!

c) There is a real motion, but the true velocity is less
than the velocity of light, and the apparent large velocity is
due some illusionary effect.

It is generally believed that some variation of the latter
effect is the correct explanation, and one of the most attractive
explanations is based on the relativistic motion model of Rees.

If, in fact, there are two components separating with a
highly relativistic velocity, then the effect of the finite
travel time of the signals from the two components will give the
illusion of a transverse expansion velocity, V, given by (v is
the true velocity in the rest frame of the source and θ is the
angle between the motion and the line of sight)

$$V = \frac{v \sin \theta}{1 - (v/c) \cos \theta} \tag{4.1}$$

This is maximum when $\sin^{-1}\theta \sim \gamma^{-1}$, with $\gamma = (1 - v^2/c^2)^{-\frac{1}{2}}$. Then

$$V_{max} = \gamma c.$$

For apparent velocities $V \sim 5c$, $\gamma \sim 5$, and $\theta \sim 10°$.

There are, however, several difficulties in accepting this
model. First, the differential Doppler shift should cause the
luminosity of the approaching component to exceed that of the
receding component by a factor of (α is the spectral index)

$$\frac{S_{app}}{S_{rec}} = \left[\frac{(1 + v/c \cos \theta)}{(1 - v/c \cos \theta)} \right]^{3-\alpha} \tag{4.2}$$

For $\gamma \sim 5$, the receding component should be so faint that
in general it is not observable. However, there may in fact be
numerous components which are ejected more or less isotropically.
The ones approaching near to the line of sight then appear very
much more intense than the others which are then not detected

in the model fitting procedures used to describe the source structure.

A further objection is that it takes considerable energy, $E \sim \gamma\ mc^2$, to accelerate material to relativistic velocities. Moreover, even if this velocity occurs initially, interaction with the surrounding medium will quickly decelerate the motion (see the lectures by Rees). This is especially true if the motion occurs in the dense nucleus of a galaxy such as 3C 120 or within a quasar. In 3C 273 the dimensions of the triple radio source is less than the estimated size of the dense emission line region. But of course we do not know that the radio and optical emission coincide, since the position agreement is accurate only to $\sim 0\rlap{.}''05$ while the dimensions are $< 0\rlap{.}''01$. Also it is possible that a previous outburst has swept out a vacuum, allowing the expansion to continue with negligible deceleration. Nor is it necessary that there be actual physical motion. Instead, an electromagnetic beam propagating out from the center may "illuminate" a cloud. This gives the kinematic illusion of faster-than-light motion without the problems associated with the differential Doppler shift or deceleration (see discussion by Rees).

In any event an explanation based on the light travel time from relativistic components requires that the motion always be nearly along the line of sight. We have already noted that in the compact components of extended sources, the compact components appear aligned with the extended ones, which are assumed to be randomly oriented. If apparent faster-than-light motion is observed in a number of these objects, then the above model is unlikely to be valid.

REFERENCES

Cohen, M. H., 1969, Ann. Rev. of Astron. and Astrophys. 7, 619.
Cohen, M. H., Moffet, A. T., Romney, J. D., Schilizzi, R. T., Shaffer, D., Kellermann, K. I., Purcell, G. H., Grove, G., Swenson, G. W., Yen, J. L., Rinehart, R., Pauliny-Toth, I. I. K., Preuss, E., Witzel, A., and Graham, D., 1975, Astrophys. J. 301, 249.
Condon, J., and Dressel, L., 1974, Astrophys. Letters 15, 203.
Dent, W. A., and Hobbs, R. W., 1973, Astron. J. 78, 163.
Dent, W. A., and Kojoian, G., 1972, Astron. J. 77, 819.
Fomalont, E. B., and Wright, M. C. H., 1974, Galactic and Extragalactic Radio Astronomy, Kellermann and Verschuur, ed., Springer-Verlag.
Kellermann, K. I., and Pauliny-Toth, I. I. K., 1968, Ann. Rev. Astron. and Astrophys. 6, 417.
Kellermann, K. I., 1974, Galactic and Extragalactic Radio Astronomy, Kellermann and Verschuur, ed., Springer-Verlag.

Kellermann, K. I., Clark, B. G., Niell, A. E., and Shaffer, D. B. 1975, Astrophys. J. 197, L113.

Kellermann, K. I., Shaffer, D. B., Purcell, G. H., Pauliny-Toth, I. I. K., Preuss, E., Witzel, A., Schilizzi, R. T., Cohen, M. H., Moffet, A. T., Romney, J. D., Niell, A. E., Rinehart, R., 1976, Astrophys. L., in press.

Medd, W. J., Andrew, B. H., Harvey, G. A., and Locke, J., 1972, Mem. Roy. Astron. Soc. 77, 109.

Pauliny-Toth, I. I. K., Preuss, E., Witzel, A., Kellermann, K. I., Shaffer, D. B., Purcell, G., Grove, G., Jones, D., Cohen, M. H., Schilizzi, R. T., Romney, Moffet, A. T., and Rinehart, R., 1975, Nature, in press.

Rees, M., 1967, Mon. Not. Roy. Astron. Soc. 135, 341.

Schilizzi, R. T., Cohen, M. H., Romney, J., Shaffer, D., Kellermann, K. I., Swenson, G. W., Yen, J. L., Rinehart, R., 1975, Astrophys. J. 301, 263.

Seielstad, G. A., 1974, Astrophys. J. 193, 55.

Shaffer, D. B., Cohen, M. H., Romney, J. D., Schilizzi, R. T., Kellermann, K. I., Swenson, G. W., Yen, J. L., Rinehart, R., 1975, Astrophys. J. 301, 256.

Wittels, J. J., Knight, C. A., Shapiro, I. I., Hinteregger, H. F., Rogers, A. E. E., Whitney, A. R., Clark, T. A., Hutton, L. K., Marandino, G. E., Niell, A. E., Ronnang, B., Rydbeck, O. E. H., Klemperer, W. K., and Warnock, W. W., 1975, Astrophys. J. 196, 13.

THE OPTICAL PROPERTIES OF RADIO GALAXIES

E. M. Burbidge

Department of Physics
University of California, San Diego
La Jolla, California 92093, U. S. A.

In these two lectures I shall describe the optical properties of
galaxies that are identified with radio sources, both from the
morphological and structural and from the spectroscopic and
spectrophotometric points of view. In Section 1 I shall discuss
the kinds of galaxies which are found to be radio sources. The
types of spectra observed will be discussed in Section 2. In
Section 3 we shall inquire into what kind of evidence can be
found to indicate that a particular galaxy is a candidate for identification with a radio source, what evidence exists for the
occurrence of a "violent event" in such a galaxy, and what can
be found out about a radio galaxy from optical studies.

In Section 4 I shall describe some individual objects that
provide illustrations for the various aspects that are described
in general terms in the preceding sections, and then finally in
Section 5 we shall take a look at radio galaxies as a cosmological
tool, as seen from the optical point of view.

1.   MORPHOLOGICAL CHARACTER OF RADIO GALAXIES

It was realized soon after the first identifications of radio
galaxies were made that many of them fell into the category of
giant elliptical galaxies (E). In the meanwhile, two new morpho-

G. Setti (ed.), The Physics of Non-Thermal Radio Sources, 41–53. All Rights Reserved.
Copyright © 1976 by D. Reidel Publishing Company, Dordrecht-Holland.

logical classes for galaxies in general, which were additional
to the Hubble classes, were set up by Morgan (1958). These
were the D and the N classes. Since the N class has come to
be used very widely for compact radio galaxies, it is worthwhile to recall Morgan's original definition: "Systems having
small, brilliant nuclei superposed on a considerably fainter
background."

By 1964, when a considerable number of identifications
were available, Matthews, Morgan, and Schmidt (1964) listed
morphological classes for 48 radio galaxies (their list of 52
objects included 4 QSOs). The majority of these objects were
identifications for 3C radio sources, i.e. strong sources. In
their paper a more detailed description of the D galaxies was
given: they were galaxies having an elliptical-like nucleus
surrounded by an extensive envelope; thus they were now clearly
distinguished from Hubble's SO class (whereas the distinction
had not been clearcut in Morgan's 1958 classification). D
galaxies were now defined to be objects that were not highly
flattened, while the Hubble-Sandage class SO applied to flattened
systems. Matthews et al. (1964) introduced a new subdivision
of class D which has since proved to be very important: very
large D-type galaxies in clusters, having diameters 3-4 times
those of average D and E galaxies were called cD, a notation
taken over from stellar spectroscopy where "c" denoted
"supergiant."

Of the 48 radio galaxies in the list of Matthews et al., 21
were of type D or E, and 10 were of type cD. Thus these by
far comprised the bulk of the identifications. There were 6
spirals, 2 irregulars, and 5 of a new class named "dumbbell,"
described as being allied to the D galaxies, in which two
separated, approximately equal, nuclei were embedded in a
common envelope. There were only 4 N-type objects, described
there as having brilliant starlike nuclei containing most of the
luminosity of the system, and having a faint, nebulous envelope
of small visible extent. These results from Matthews et al. are
tabulated below.

Table I

D or E	21	I	2
cD	10	db	5
S	6	N	4

These categories have persisted and still serve to
describe the types of galaxies found to be radio emitters. D,
E, or cD still comprise the majority of radio galaxies, and
irregulars are quite rare. Seyfert galaxies belong in both
classes S and N, because some of them, like NGC 1068, have
distinct and normal-seeming spiral structure while others,
like NGC 4151 and 3C 120, have chaotic or ill-defined and
faint knotty structure. All Seyferts, however, satisfy Morgan's
original definition of N-type in that they have an intensely
bright starlike nucleus.

A final point to be made before we leave this discussion
of the morphology of radio galaxies is that the strongest radio
sources tend to be the D, E, or cD objects and the N systems .
Normal spirals and irregulars are much weaker, and in spirals
the radio emission is often confined to the nuclear region.
Some Seyferts, however, like NGC 4151, are quite weak radio
emitters. Radio galaxies of D, E, or cD type are often the
brightest members in clusters of galaxies, and are massive
objects.

2.   SPECTRA OF RADIO GALAXIES

Since galaxies in general are composed of stars, gas, and dust,
we expect the spectra of radio galaxies to demonstrate the
presence of these components in various degrees of importance.
For optical spectra, this means we expect to see spectral
features arising in stars and/or gas, as we do in normal
galaxies. But there is an additional component which is certainly present in radio galaxies in order to produce the radio
emission -- that is, the high-energy charged particles and
magnetic fields which produce synchrotron radiation. If the
particle energies are high enough, they can produce synchrotron radiation at optical (and even at X-ray wavelengths) and
this means there may be non-thermal continuous radiation which
may dominate in the optical part of the spectrum.

However, a large number of the classical radio galaxies
have most of their radio emission in extended regions outside
the galaxy, occurring usually in two lobes on either side of the
optical object or in extended tails or halos. When there are
additional compact radio components centered on the galaxies
themselves, these tend to be weaker than the outer extended
emission. Thus in those galaxies where the radio emission is

extended, we do not tend to find compact non-thermal emission
at optical wavelengths although there are exceptions to this,
and we shall expect to see an optical spectrum characteristic
of the galaxy itself.   This will in general consist of absorption
features, e. g. Ca II H and K, the G-band from CH, the Mg "b"
feature, and Na I D lines, i. e. all those features characteristic
of the relatively cool normal stars which are emitting most of
the light of the galaxy.   This is indeed what is found -- such
absorption features are characteristically found in the E, D,
and cD radio galaxies, and the optical continua of such radio
galaxies are like those of normal E galaxies.

The situation is different for the N-systems.   Here the
spectrum usually shows strong emission lines, clearly coming
from hot gas, and there is usually no indication of a spectral
component due to stars.   In Cyg A, there is extended luminosity
which has been assumed to be a cD galaxy but there is no
absorption-line spectrum and only a very strong emission-line
spectrum coming from hot gas (this object is discussed in detail
in Section 4).

A useful way to categorize the radio galaxies is by the
type of optical spectrum that they have -- this can be done even
for faint objects.   Burbidge and O'Dell (1973) analyzed the redshifts for 222 radio galaxies, of which 152 had no strong
emission lines in their spectra, 36 had strong emission lines
but were not compact, and 34 had strong emission and were
compact.   The data were obtained from Moffet (1974), Sargent
(1973), Demoulin (1970), and Spinrad and Smith (1973).   Burbidge
(1970) tabulated data for 92 radio galaxies, of which about half
had strong emission lines.   It must always be remembered that
unless we can see a characteristic stellar-type spectrum, such
as would be produced by an aggregate of stars of the spectral
types characteristically found to emit most of the light in normal
galaxies, then we have no direct evidence that normal stars are
present, and in some of the highly compact N systems, they may
not be.

3.   CRITERIA FOR ADOPTING A GALAXY AS IDENTIFI-
        CATION FOR A RADIO SOURCE; EVIDENCE FOR
        "ACTIVITY" IN SUCH GALAXIES

Much has been written about problems of understanding the very
large energy release required to explain the strong radio sources,
and the emergence of this energy in the form of high-energy

charged particles and magnetic fields. The only feasible
energy source appears to be gravitational energy release,
accompanying or followed by an explosive event. What will be
the evidence for this?

Firstly, the most direct evidence is the presence of the
non-thermal radiation itself, either at radio, infrared, optical,
or X-ray wavelengths. At optical wavelengths, non-thermal
radiation may be evidenced by a continuum spectrum having a
slope of power-law form:

$$F(\nu) = \nu^{-\alpha} \qquad .$$

Variable polarization may be present. The flux may be
variable, often on a short timescale. All these are indications
of the presence of non-thermal radiation, i. e. of high-energy
particles.

If an explosive event has occurred, accompanying the
collapse of a considerable mass of material to very high density,
we may argue by analogy with the near-at-hand and well-known
example, the Crab Nebula. Non-thermal radiation at all wavelengths is present, so is a high degree of polarization, so is a
collapsed super-dense object (the pulsar or neutron star).

What are the other manifestations of the energy release in
the Crab? They are: large-scale highly supersonic motions in
the gas, and ionization and excitation of that gas. We can make
a direct analogy and say that in a galaxy which has undergone a
"violent event," one would expect to see large-scale gas motions
that are supersonic and that are clearly separable from the
normal rotational motion of the galaxy. Turning this the other
way around, if we see non-circular motions in the gas of a
galaxy, this will provide support for its identification with a
radio source. Similarly, a strong emission-line spectrum and
the presence of lines of highly ionized atoms will also support
such an identification.

A third indicator of activity in galaxies is the presence of
unusual structure -- jets, plumes, disordered or peculiar dust
features, and filamentary structure or extensions which cause
the galaxy to be labeled "peculiar."

Apart from the presence of clearly discernible non-thermal
radiation at wavelengths in addition to the radio emission,
however, the existence of the other indicators, i. e. high ionization and excitation in hot gas, distinctive non-circular motions,
and peculiar structure, do not provide firm evidence that a

galaxy possessing these features is a radio galaxy. In the
early history of identification work on radio galaxies, when radio
positions had large errors, there were some mis-identifications
with galaxies having strong emission lines in their spectra,
simply because the first of the more distant radio galaxies to
be identified, Cygnus A, does indeed have a very strong
emission-line spectrum (we shall return to discuss this object
in detail later). It has been established now that many galaxies
do have strong emission lines but are not radio emitters,
except perhaps at a very low level.

Even more important: unusual structure is not at all a
definitive indicator for a radio galaxy. Some galaxies have
quite normal morphological appearance, yet are strong radio
sources. Conversely, some galaxies have unusual structure
but are not radio galaxies. In fact, we do not have any good
data on the proportion of normal (non-radio) galaxies which
have any structural peculiarities such as extended plumes,
jets, etc.

Let us now turn to the question of what information can
be yielded by optical studies of a radio galaxy. First and
foremost, of course, is the redshift. It has been an unfortunate
inherent aspect of radio astronomy that one cannot get a
redshift -- most radio galaxies do not seem to have sufficient
cold H I to give a 21-cm line, and in any case, to look for such
a line one must tune the receiver to an appropriate frequency
and hence must know the redshift, at least approximately, in
advance.

The optical observations one wishes to make for a galaxy
suggested to be a radio source identification are:

a.   Use the Palomar Sky Survey, or other Schmidt
plates in the Southern Hemisphere, to determine the morphological type.

b.   Obtain a low-dispersion spectrum to measure the
redshift and to determine the character of the spectrum, i. e.
whether it has only absorption lines, or modest emission such
as [O II] $\lambda 3727$ and maybe $H\alpha$, or strong emission lines.

c.   In many cases, it will be useful to obtain direct
photographs with larger scale, and maybe to obtain both short
exposures to show structure in the nucleus and really "deep"
plates to show faint outer structure. Calibrated plates or scans
across the galaxy will give the distribution of stars in E, D, or
cD galaxies. Special narrow-band filter photographs may also
be useful. We shall see examples of information yielded in
these ways in the next section.

      d.   If the object has only absorption lines, it will be useful to analyze the composite spectrum in order to see what are the population and range of spectral types of the stars making up the galaxy.

      e.   If the object has emission lines, spectrophotometric study of them with good signal-to-noise will enable the physical conditions, composition, and source of ionization of the gas to be studied, both in the inner and outer regions. Spectrophotometry will also give the power-law index of the continuum if non-thermal optical emission is present.

      f.   Kinematic studies of both the stars and gas can be made from Doppler shifts measured on long-slit spectrograms. The mass of the galaxy can be studied, also the nature of any non-circular motions which are detected.

      g.   Special studies may be suggested by unusual features of the galaxy or of its spectrum -- e. g., the jet in M87 (see Section 4) has obviously warranted much study, as do the structures recently discovered in Cen A. The narrow absorption lines, indicative of outflowing gas from the nuclear region, found in NGC 4151 and Markarian 231, are also unusual and very important to study.

      h.   Finally, polarization should be looked for, both in lines and in the continuum.

      To sum up, we can study:

Direct Photographs
      Morphology
      Distribution of light
      Presence of unusual structural features

Spectra
      Redshift
      Type of spectrum
      Type of stars in galaxy
      Character of non-thermal emission if present
      Physical conditions in excited gas
      Outflow of gas if present
      Kinematics of object

Polarimetric
      Polarization in continuum
      Polarization in lines

With these tools, the real aim is to find out:

      a.   The nature of galaxies that become strong radio emitters,

      b.   What is occurring during and after a "violent event."

4.   OPTICAL FEATURES OF SOME REPRESENTATIVE GALAXIES

A selection of radio galaxies, necessarily limited, will be described in order to illustrate the points made in the preceding sections. We shall consider the examples galaxy by galaxy; some galaxies will be interesting by virtue of several of the optical properties described above, while some will illustrate only one of these features.

4.1   NGC 4782-3 = 3C 278

This is a "dumbbell" or double elliptical in a common envelope. It lies in a cluster. Both components of the "dumbbell" show only absorption lines in their spectra (Page 1952; Greenstein 1961); the velocity difference, presumably due wholly to orbital motion, is large and implies that the galaxies are massive ($\sim 10^{12}$ $M_\odot$). The radio source cannot be located in one or other of the pair, so we do not know which is responsible for the "violent event." The interesting feature of the system, however, is that the distribution of stars (whose integrated spectra indicate a composition of stars normal for giant E galaxies) are strikingly different in the central regions of the two galaxies (Burbidge, Burbidge, and Crampin 1964). NGC 4783 has a strong central concentration, like other galaxies lying near it in the field, while NGC 4782 has a diffuse stellar distribution in the central region. One has the impression that "something happened" to the normal high degree of concentration in NGC 4782. Unfortunately, however, we still have little data on the distribution of stars in the central regions of normal non-radio E galaxies, thus we do not know whether this type of structure is actually highly abnormal or not.

4.2   NGC 5128 = Centaurus A

The striking and very obvious structural feature of this galaxy is the broad, chaotic dust lane across the central region of what otherwise would seem to have most of the characteristics of a giant E galaxy. One wonders: was the dust produced in one or more violent events? Or was it present before, in amounts that perhaps contributed toward making the galaxy susceptible to violent events, and was it then disturbed and pushed outward by such events? That there has been more than one "violent event" is evident from the complex structure of the radio emission in Cen A.

The presence of an outer plume-like structure in NGC 5128, elongated along the rotation axis, more or less in the direction of elongation of the radio source, was discovered several years ago by Hugh Johnson. The recent discovery of interesting structures aligned in the same general direction has been described elsewhere in these lectures. This galaxy is a fine example of non-thermal emission -- from its several radio components to its X-ray component which appears to be located in the center of the galaxy.

The optical spectrum does not exhibit strong emission lines, but $H\alpha$ and low-ionization emission lines characteristic of many normal galaxies are found in and near the dust lane. Non-circular, as well as rotational, motions are present (Burbidge and Burbidge 1959). An interesting hint that the stellar component may be unusual has come recently from unpublished observations made at the Anglo-Australian Observatory and communicated by J. Wampler. The absorption-line spectrum appears to suggest that the stars are weak-lined (i. e., metal-poor). We await further detailed optical studies of this very interesting radio galaxy.

4. 3   M87 = Vir A

The optical jet emerging from the nucleus of M87 is too well known to dwell on here. We should simply point out that, despite considerable effort, no emission lines have been detected in the jet, which has a pure non-thermal, highly polarized, optical continuum with a flat spectrum, giving it a blue color. The blobs in the jet do seem, from recent studies, to have very compact structure. The stars in the galaxy yield a normal composite spectrum characteristic of a giant E galaxy. Weak [O II] $\lambda3727$ is present in the very nucleus, but the amount of gas must be small and the excitation is apparently low.

4.4   MCG 9-13-66 = DA 240

This is an E galaxy identified with a radio source of unusually
large extent (Willis, Strom, and Wilson 1974), with a redshift
z = 0.035. Direct photography at Kitt Peak National Observatory
revealed that there is a jet-like outer structure aligned some
62° to the direction of extent of the outer radio lobes (Burbidge,
Smith, and Burbidge 1975). Weak emission lines of $H\alpha$, [N II],
and [O II] are seen in the nucleus, but have not been detected in
the jet. The jet is bluer than the nuclear region, but is certainly
not as blue as the jet in M87 (It has B - V = 0.86, U - B = 0.66
as compared to B - V = 1.19, U - B = 1.15 in the nuclear region;
these measures were made at Lick on the Lick scanner photo-
metric system). Normal absorption features characteristic of
the usual stellar population for a giant E galaxy are seen (i.e.,
Ca II, H and K, the Mg b band, and the Na I D lines).

4.5   Cyg A

Since this is such an important radio source, upon which so
many theoretical deductions have been based, we will spend
some time discussing its optical properties. There have been
two recent spectrophotometric studies which have given much
new information on the excited hot gas (Mitton and Mitton 1972;
Osterbrock and Miller 1975).
        Firstly, it is well to emphasize that this object, often
used as a prototype in drawing cosmological deductions from
radio sources, is quite atypical. It has a very strong emission-
line spectrum, exhibiting a wide range of ionization -- lines are
seen ranging from [O I] through [Ne V] all the way to [Fe X]
(probably, although not certainly, present). Although they are
so strong, the emission lines are not nearly as broad as those
characteristic of the nuclei of Seyfert galaxies; they have full-
width-half-maximum = 500 km/sec ± 100 km/sec.
        There is extended luminosity around the brightest part of
Cyg A, which lies near the galactic plane and is heavily reddened.
It is always assumed that this outer luminosity is a giant E or D
galaxy, but it is well to note that the high signal-to-noise spectro-
photometry of Osterbrock and Miller yielded no sign of absorption lines from a stellar component, and thus there is no evidence
for the presence of stars in this object. After correction for the
large reddening and extinction (E_{B-V} = 0.69 mag, determined
from measuring the Balmer decrement), the absolute visual
magnitude derived was M_V = -24.9; this was essentially all from

emission lines plus a non-thermal continuum having a slope
(corrected for reddening) given by

$$F_\nu \propto \nu^{-1.6 \pm 0.2} \qquad .$$

Physical parameters in the gas are: a mean electron
temperature T_e of 10,000-15,000 °K and a mean electron
density $N_e = 8 \times 10^2$ cm^{-3} (note that this is much lower than
N_e derived from those QSOs which have been analyzed, where
N_e is characteristically 10^6 cm^{-3} and may be as high as 10^8
cm^{-3}).

The gas in Cyg A is clearly very filamentary and non-
homogeneous -- much more so than in typical planetary nebulae
and H II regions in our Galaxy. The mass of hot gas required
to give the observed flux in Hβ is 2×10^7 M$_\odot$, and the observed
extent of the emission lines demonstrate that this excited gas
must be distributed throughout the volume with a very small
filling factor.

The energy mechanism for heating the gas might be shock-
wave heating or photoionization; Osterbrock and Miller believed
the latter to be more likely, with the UV ionizing flux coming
from synchrotron emission.

4.6   Radio Galaxies with Interesting Velocity Fields

There is not time to discuss examples in detail, so we mention
a few striking cases. 3C 390.3, an N-type galaxy, has enormously wide emission lines extending some 17,000 km/sec, and
has two components separated by some 4000 km/sec (Burbidge
and Burbidge 1971).

Seyfert galaxies typically have very broad strong emission
lines in their nuclei, and NGC 1275 = Per A has, in addition,
the well-known filamentary structure with a velocity 3000 km/sec
greater than that of the main body of the galaxy (Minkowski 1957;
Burbidge and Burbidge 1965). The stars in the galaxy NGC 1275
are of earlier spectral type than those characteristic of E
galaxies. A careful search in the gas with the displaced velocity
component has failed to show any sign of a "nucleus" for this
gas, thus disposing finally of the original suggestion that we
were seeing two galaxies in collision. Thus we are apparently
seeing gas being ejected from the nucleus of NGC 1275 at some
3000 km/sec. Understanding the geometry of this process has
been further complicated by the discovery of 21-cm absorption

at the <u>displaced</u> velocity in front of a very compact radio component in the nucleus of the galaxy. Finally, a narrow-band
interference filter photograph in Hα at the velocity of the
undisplaced gas by Lynds (1970) has shown a very interesting
filamentary structure threading through the galaxy, and looking
remarkably like the filamentary structure in the Crab Nebula!

5.   RADIO GALAXIES AS A COSMOLOGICAL TOOL

There is time only for a very brief discussion of the use of
radio galaxies as a cosmological tool. Since we are considering
only the optical properties of radio galaxies, the traditional use
of the log N - log S relation, where S is the radio flux, will not
be discussed.

The obvious cosmological use that can be made of radio
galaxies is to plot them in a Hubble diagram, with log (redshift)
plotted against magnitude. Since one needs to choose galaxies
that resemble "standard candles" as nearly as possible, a well-
defined sample of galaxies is needed, and the strongest radio
sources are the obvious candidates. This means the 3C galaxies.
Excluding the QSOs, there are 253 3C sources out of the
Galactic plane which are considered to be extragalactic. At the
latest count, 140 have been identified with galaxies; some are
very faint and at the limit of the Palomar 200-inch or Kitt Peak
4-m telescopes. There are now about 100 measured redshifts
among this sample. Log N - log S plots for the identified
galaxies and for those with redshifts (Burbidge and Narlikar
1975) do not indicate any cosmological effects -- the former
yield a slope of -1.5, and the latter one of -1.1, i.e. much
flatter slopes than that usually quoted by radio astronomers.

The Hubble diagram (Smith and Spinrad, private communication) shows considerable scatter. In this it resembles the
plot given by Gunn and Oke (1975) for non-radio galaxies that
are the brightest members of clusters, but the radio galaxies
tend to lie slightly <u>above</u> Gunn and Oke's plot, meaning that the
radio galaxies tend to be somewhat brighter than non-radio
galaxies. This, of course, is true in particular of 3C 295 which
Gunn and Oke found lay well above their mean plot.

It is well to emphasize that the radio galaxies which Smith
and Spinrad considered all fall into the E, D, or cD category (as
far as can be determined for such faint images). Some N
galaxies for which they also measured redshifts would fall well
off the mean Hubble plot, since they have large redshifts lying
in the domain occupied by the QSOs.

What is lacking at present for such a Hubble plot is accurate photometry. Even if this is done (involving considerable effort at the telescope) there will be the additional worry about luminosity evolution in elliptical galaxies. Although doing the photometry at a fixed wavelength band in the rest-wavelength system is now feasible, by using the new scanner spectrophotometers at the various observatories, this problem of the intrinsic luminosity remains, to my mind, such a serious obstacle that I am quite doubtful whether the deceleration parameter q_0 will ever be pinned down from optical studies of radio galaxies.

REFERENCES

Burbidge, E. M., and Burbidge, G. R., 1959, Ap. J. 129, 271.
Burbidge, E. M., and Burbidge, G. R., 1965, Ap. J. 142, 1351.
Burbidge, E. M., and Burbidge, G. R., 1971, Ap. J. 163, 121.
Burbidge, E. M., Burbidge, G. R., and Crampin, D. J., 1964, Ap. J. 140, 1462.
Burbidge, E. M., Smith, H. E., and Burbidge, G. R., 1975, Ap. J. (Letters) 199, L137.
Burbidge, G. R., 1970, Ann. Rev. Astron. & Ap. 8, 369.
Burbidge, G. R., and Narlikar, J. V., 1975, Ap. J., in press.
Burbidge, G. R., and O'Dell, S. L., 1973, Ap. J. (Letters) 186, L59.
Demoulin, M.-H., 1970, Ap. J. (Letters) 160, L79.
Greenstein, J. L., 1961, Ap. J. 133, 335.
Gunn, J. E., and Oke, J. B., 1975, Ap. J. 195, 255.
Lynds, C. R., 1970, Ap. J. (Letters) 159, L151.
Matthews, T. A., Morgan, W. W., and Schmidt, M., 1964, Ap. J. 140, 35.
Minkowski, R., 1957, in Radio Astronomy, I. A. U. Symp. No. 4 (ed. H. C. van de Hulst; Cambridge Univ. Press), p. 107.
Mitton, S., and Mitton, J., 1972, M. N. R. A. S. 158, 245.
Moffett, A. T., 1974, Stars and Stellar Systems, Vol. 9, in press.
Morgan, W. W., 1958, Publ. Astron. Soc. Pacific 70, 364.
Osterbrock, D. E., and Miller, J. S., 1975, Ap. J. 197, 535.
Page, T. L., 1952, Ap. J. 116, 63.
Sargent, W. L. W., 1973, Ap. J. (Letters) 182, L13.
Spinrad, H., and Smith, H. E., 1973, Ap. J. (Letters) 179, L71.
Willis, A. G., Strom, R. C., and Wilson, A. S., 1974, Nature 250, 625.

OPTICAL STUDIES OF QUASARS AND BL LAC OBJECTS

P.A. Strittmatter

Steward Observatory, University of Arizona
Tucson, Arizona, USA

1. HISTORICAL INTRODUCTION

Quasistellar objects must count as one of the great discoveries
of radio astronomy. In the late 1950's, as radio source positions
became more accurate, a number of blue stellar objects were
identified with radio sources. The first was 3C 48 which was
identified with a 16^m object by Mathews and Sandage; it was also
found to have an unusual spectrum. Several other sources of this
type were located, found to be variable and deduced on that basis
to be local objects, i.e. within our galaxy. Following the classical lunar occultation measurement of 3C 273 by Hazard, Mackey and
Shimmins (1963) a spectrum was obtained by Schmidt (1963) who, in
recognizing a Balmer series spectrum shifted by a redshift $z = \Delta\lambda/\lambda_0 =$
$= 0.158$, provided the breakthrough in interpreting the spectra of
these objects. Other redshifts followed in quick succession, 3C 48
turning out to have a redshift $z = 0.367$ and 3C 9 yielding the unprecedented value of $z = 2.02$. Almost immediately the QSOs were
transformed from variable local radio sources to the most luminous
objects in the universe. With this realization came great hopes of
using the QSOs as probes of the universe out to redshifts where
non-linear geometrical effects are large. Again the lesson had to
be relearned that one needs to understand one's probes before using
them to derive astronomical information. In these lectures I shall
therefore be dealing with the objects themselves rather than what
they may be able to tell us about the geometry of space-time.
    This early and rapid development in our knowledge of QSOs amply
illustrates the central problem that has occupied much of the intervening time, namely that of reconciling the short time variability
with the high redshift. The consensus today is that the redshift

G. Setti (ed.), The Physics of Non-Thermal Radio Sources, 55–66. All Rights Reserved.
Copyright © 1976 by D. Reidel Publishing Company, Dordrecht-Holland.

is due to Hubble expansion but the variability has not yet been
adequately explained. Some astronomers, however, continue to
seek alternative explanations of the redshift. We will assume in
most of what follows that the redshift is indeed due to Hubble's
Law.

2. OPTICAL PROPERTIES OF QSOs

Optical astronomy has been concerned in finding QSOs, both as radio
source counterparts and directly, and obtaining spectra. The latter
may then be used for cosmological purposes or attempts to understand
the objects themselves. In fact, the optical astronomer in studying
both emission and absorption line data has accumulated a substantial
body of data on essentially thermal properties of QSOs. Only the
continuum studies, especially data on variations, have provided
information on presumably relativistic processes. In this respect,
optical astronomy has little advantage over other spectral regions
except that it produces redshifts and hence, using Hubble's Law,
luminosity data. Most of the interpretations of optical data have
concentrated on the wealth of thermal information. Before proceeding
with further discussion, however, I think a summary of properties
would be useful. In this context, I note the existence of a second
class of object which many of us believe is closely related to the
QSOs and which also merits inclusion here - the BL Lac objects.
The properties of typical QSOs and BL Lac objects are summarized in
Table 1. These we now deal with in turn.

2.1 Optical Appearance

Most QSOs appear stellar on sky limited photographs - hence
their name. Nonetheless, a substantial fraction of low z (< 0.4)
QSOs show structure (e.g. 3C 48) which Kristian has suggested is
due to a surrounding elliptical galaxy. This view is consistent
with the absence of such objects at higher redshifts. On the other
hand, in the only QSO for which a spectroscopic study of the nebula
has been carried out, 3C 48, the spectrum is dominated by emission
lines, [OIII] being especially strong (Wampler, Robinson, Baldwin
and Burbidge 1975). In 3C 273 the optical extension is really a
jet, also unlike an E galaxy but perhaps related to the well known
jet in M 87. The nature of the fuzz around QSOs thus remains unclear.
For the BL Lac objects the situation is somewhat clearer. The
majority of these objects appear to be surrounded by nebulosities and
in some cases quite clearly these are elliptical galaxies (e.g. AP
Lib[†] and 1101+38). Some (e.g. OJ 287, 0735+178) appear to be stellar.

[†]During the summer school I indicated doubts as to the redshift
(z = 0.048) of AP Lib determined by Disney, Peterson and Rogers 1974;
the result has been confirmed subsequently.

TABLE 1

Properties	QSOs	BL Lac Objects
Optical Appearance	Stellar or slightly extended at low red-shift	Usually extended, sometimes stellar
Color	Usually "blue" on PSS	Usually "neutral" on PSS
Variability	Normally modest $\Delta m \leqslant 0.2^m$ Occasionally large (e.g. 3C 446, $\Delta m \sim 4^m$) $\Delta t \sim$ weeks	All variable $\Delta m > 0.3^m$ $\Delta t \sim$ days
Polarization	Normally modest, < 2 percent Occasionally large (e.g. 3C 446 $\sim$ 10%) variable $\Delta t \sim$ days	All large and variable $p \sim 20\%$ $\Delta t \leqslant$ day
Continuum	Power law $\alpha \sim 1$ typically, but $\alpha \sim 0$ (e.g. 3C 286) and $\alpha \sim 2$ (e.g. 3C 446); Balmer jump seen sometimes Evidence of curvature	Power law $\alpha \sim 2-3$ typically Evidence of curvature
Emission Lines	Resonance, broad (e.g. Ly-α, CIV) Forbidden, usually sharp (e.g. OIII) Recombination, usually broad (H, HeI, HeII) Intercombination, broad (e.g. CIII) Lines often asymmetrical Equivalent width up to 2000A (Ly-α) Redshifts $0.1 \leqslant z \leqslant 3.53$	None (Definition)
Absorption Lines	Few (if any), $z_{em} < 1.0$ Many, $z_{em} \geqslant 2.2$ Resonance lines usually Generally sharp $\Delta v \leqslant 10$ km/sec $z_{abs} \leqslant z_{em}$ usually $z_{abs} > z_{em}$ occasionally (e.g. Ton 1530) Occasional P Cygni profiles (PHL 5200)	One object only so far, 0735+178, MgII ($z_{abs} = 0.424$)

In those cases where the galaxy redshift has been determined it is
inevitably small. BL Lac itself is extended and the nature of the
fuzz is disputed. According to Oke and Gunn (1974) the spectrum
is similar to that of an E galaxy at a redshift of z = 0.07; this
result was not confirmed by Baldwin et al. (1975) so that in the
case of BL Lac itself the nature of the nebulosity remains in
doubt.

2.2 Color

QSOs generally appear blue on the Palomar Sky Survey - that
is the image appears stronger relative to the average star on the
blue as compared to the red plate. This property was of great
value in the early days of radio source identifications since blue
stars are comparatively rare. Some QSOs (e.g. OH 471, 0938+119),
however, appear neutral or even red. The BL Lac objects appear
neutral, a fact which may well have lowered their chance of identification in earlier (less accurate) radio surveys.

2.3 Variability

Rather little attention has been paid in recent years to
studies of the optical variability of QSOs, perhaps because they
are rarely dramatic. It should, however, be reemphasized that it
was this property which led to their classification as local objects.
Typical variations are of an order of $0.1 - 0.2^m$ over time scales
of a week. Some QSOs (e.g. 3C 345, 3C 446) have in the past varied
dramatically (Kinman 1975 and references therein). There is little
evidence, however, of any changes in the emission line strengths
of QSOs during such outbursts. The BL Lac objects are all known
to vary substantially. Random variation of $\gtrsim 0.3^m$ on time scales
of a day or so are typical. Long term variations with amplitudes
of several magnitudes are, however, well documented in certain
objects. For example, OJ 287 reached $\sim 12^m$ in 1974 and has now
declined to $\sim 15^m$.

2.4 Polarization

Most QSOs are only modestly polarized if at all in the optical.
Once again, however, there are certain exceptions of which 3C 345
and 3C 446 are again examples (Kinman 1975 for a discussion).
Both objects have shown polarizations which exceed 10% and vary
rapidly on occasion from night to night. The BL Lac objects are
all strongly and variably polarized, 0738+178 having on one occasion
reached 30% (Carswell et al. 1974). Time scales of variations in
both strength and position angle of polarization can be shorter
than one night. The question of possible polarization in emission
lines requires further observational sutdy.

2.5 The Continuum Energy Distributions

In the optical and near infra-red part of the spectrum the continuous energy distributions may usually be approximated by a power law spectrum of the form $f_\nu \propto \nu^{-\alpha}$ where $0 \lesssim \alpha \lesssim 2$ (Oke, Neugebauer and Becklin 1970). For most QSOs α is in fact close to unity but 3C 286, for example, has a flat spectrum ($\alpha = 0$) while some of the more variable QSOs such as 3C 345 and 3C 446 have high indices ($\alpha \gtrsim 2$).

The BL Lac objects again have continuous energy distributions of power law form, but they are generally steeper than those found in QSOs. Typical indices are around $\alpha = 2$. The coincidence between this value and that found in the highly variable QSOs was noted by Oke et al. (1970). How or whether this is correlated with the absence of emission lines in BL Lac objects is still unknown.

2.6 Emission Lines

Emission lines in QSOs fall into three basic types, namely (i) resonance lines of common elements (e.g. Ly-α, OVI $\lambda 1033$, MgII $\lambda 2798$) (ii) recombination lines (e.g. Balmer series, HeII $\lambda 4686$) and (iii) forbidden lines among which are intercombination lines such as CIII] $\lambda 1909$ and more strongly forbidden transitions such as [OIII] $\lambda 4959$, 5007, [OII]$\lambda 3727$. The resonance lines are generally broad and asymmetric, as is CIII] $\lambda 1909$; different lines have different profiles even in the same QSO. The forbidden lines are usually relatively sharp while the recombination lines frequently show evidence for a sharp component superposed upon a broad basic structure. A very broad range of ionization appears to be present in QSOs.

It is worthwhile noting that while theoretical studies of QSOs are usually compared with a "composite average" QSO spectrum, this may be a misleading procedure as "typical" QSO spectra may evolve on a cosmological time scale. No single QSO has yet been observed over the wavelength range Ly-α - Hα. It is also a regrettable consequence of physical laws that most strongly forbidden lines are in the "visible" region of the spectrum while the resonance lines are predominantly in the UV. The different types of transitions are sensitive to different physical properties but are rarely sampled in the same QSO.

2.7 Absorption Lines

Absorption lines are a somewhat rarer phenomenon among QSOs, but when they occur, they do so in abundance. They are almost invariably due to resonance lines of common elements and as such are often found in the profile of the corresponding emission line. In general there are few claims to detection of absorption lines in low z objects although this may in part be due to the paucity of candidates. MgII $\lambda 2798$ is the only "expected" case. Attempts at

Lick and at Steward to verify the presence of MgII λ2798 absorption in the few cases where its presence was claimed have so far
proved unsuccessful. At higher redshifts, however, absorption
lines become more prominent so that for $z_{em} > 2.2$ there is no well
established case without absorption lines. For redshifts $1.8 < z <$
< 2.1 there are QSOs both with and without absorption. Again it is
worth noting that a strong observational selection arises in
observations of this type since one is confined to a limited wavelength range. It appears that, among high z_{em} QSOs, the number
density of absorption lines increases sharply shortward of Ly-α
in emission presumably due to many low optical depth clouds in
which only Ly-α is strong enough to be detected (Lynds 1971).

The absorption lines may sometimes be sorted into absorption
systems at various redshifts, usually less than the emission value.
In some QSOs (e.g. 3C 191, z_{em} = 1.95, Williams et al. 1975 and
references therein) virtually all features can be identified with
"sensible" transitions; such systems I shall call "well established".
In most high z_{em} QSOs with very complex spectra only a small fraction
of the lines can be identified in such systems. Usually there are
one or two "obvious" systems, such as system A (z = 2.309) in PHL
957 (Lowrance et al. 1972); but the rest of the "possible" systems
are usually best forgotten.

Efforts to establish the reality of redshift systems have
involved either extension of the wavelength range - usually
with negative results unless the system was already "obvious" - or
higher resolution and hence higher wavelength accuracy. This latter
approach is again usually unsuccessful in its original purpose but
has nonetheless proved interesting in that is has shown that most
"strong" lines consist of multiple sharp components (e.g. 3C 191,
Williams et al. (1975), Ton 1530, Morton and Morton (1972), PKS 0237-
-23, Boksenberg and Sargent (1975)). The effect is found down to
the current resolution limits ($\sim 10^2$ km/sec) but its origin remains
unclear.

The distribution of absorption redshifts has been the subject
of considerable study. It turns out that most absorption redshifts
are less than the corresponding emission value but tend to be
clustered within say a few hundredths thereof. There are, however,
many well established cases in which the absorption redshift is
substantially less, the most dramatic example still being PHL 938
with z_{em} = 1.95, z_{abs} = 0.61 (Burbidge, Lynds and Stockton 1968).
Even these highly shifted absorption systems turn out to be split
at higher resolution, perhaps the best example being systems A and
B in 1331+170 (z_{em} = 2.08, z_A = 1.775, z_B = 1.785; Strittmatter
et al. 1973). There are, however, some QSOs in which $z_{abs} > z_{em}$
although the excess is invariably small. In view of the asymmetries
in emission line profiles and the associated problems in determining
the emission redshifts, these cases require further study.[†]

[†]My doubts in this regard may have been excessive and have in any
case been resolved by recent studies by Weymann and Williams (1976).

While most QSO absorption lines are sharp there are some
striking exceptions to this rule. They fall into two categories,
P Cygni and radiation damping profiles. The former category is
represented by PHL 5200 and RS 23 (Scargle et al. 1970 and references therein); the emission in NV, SiIV and CIV is cut-off sharply
towards shorter wavelengths where a broad absorption trough begins.
The character of this spectrum is similar to those found in early
type stars which are intepreted as due to mass outflow. The
associated velocities are of order 10^4 km/sec in the QSOs compared
to 10^3 km/sec in the stars. The second type of broad line has been
found in PHL 957 (Lowrance et al. 1972) and in 1331+170 (Carswell
et al. 1975); it is due to Ly-α and appears to be the result of
very high column densities of neutral hydrogen ($N_H \sim 10^{21}$ cm^{-2}).
This also provides an analogy with the UV spectra of distant OB
stars which show similar absorption presumably due to interstellar
clouds (Savage and Jenkins 1972).

At the time of the summer school only one BL Lac object,
0735+178, was known to have absorption lines, namely the MgII
doublet at a redshift of z = 0.424 (Carswell et al. 1974). This
clearly established the QSO or "high redshift" nature of these objects. Subsequently AO 0235+164 has been shown to have absorption
features also (Burbidge, Caldwell, Smith, Liebert and Spinrad 1976).

3. INTERPRETATION OF THE OPTICAL DATA

3.1 The Distance Problem

Much of the discussion of QSO spectra has concentrated on the
interpretation of the redshifts. The evidence has changed little
since the early days of QSO research although positions have now
become more entrenched and more settled on a cosmological interpretation. The main problem is still in reconciling the variability
with the luminosity. The first quantitative discussion of the problem was given by Hoyle, Burbidge and Sargent (1966). Briefly the
problem is to account for the emission process, given the luminosity (from the redshift) and the dimension of the emitting region
(from the time variations). If incoherent synchrotron radiation
is responsible, it appears that strong X-ray emission would also be
expected (produced in the Compton interaction between relativistic
electrons and the observed radio photons); this is contrary to observation. To resolve this difficulty it has variously been suggested that (i) the redshift is not cosmological (i.e. the luminosity is less, (ii) the dimension is greater than that suggested by
the light time scale (due to relativistic or geometric effects), or
(iii) that some coherent radiation mechanism is involved (but which?).
A review of the current status of these arguments is given by
Jones, O'Dell and Stein (1974).

Numerous observational attempts have been made to test the cosmological interpretation. One such approach has been through absorption line data. If it could be established that an absorption arose in an intervening galaxy, this would clearly go a long way towards proving that at least a substantial part of the redshift was cosmological (Bahcall 1971). Good candidate cases do exist, for example PHL 938 (z_{abs} = 0.61, z_{em} = 1.995), 1331+170 (z_{abs} = 1.785, z_{em} = 2.08), PHL 957 (z_{abs} = 2.309, z_{em} = 2.69) in which the absorption spectra look like an interstellar absorption cloud. There is, however, no reason why such clouds could not exist in QSOs and the case of 1331+170, where the "interstellar" absorption system is coupled to a second system with z = 1.775 and a wide range of ionization, even suggests this. It should also be noticed that the only two known QSOs with radiation damping Ly-α profiles have the same velocity difference {or ratio R = $(1+z_{em})/(1+z_{abs})$} between the emission and absorption systems. Although the increase in absorption line density to higher redshifts is dramatic, it now appears that these cannot be due to intervening galaxies (or other clouds) at lower redshift since the increase in number density applies at all wavelengths down to the atmospheric cut-off. Some QSOs with redshifts z ∿ 2 (e.g. LB 8755 and BSO 6) show no absorption even in the spectral range 3200-3600A where Ly-α should be visible. The absorption line data has not so far provided an answer to the redshift question.

Other attempts have been made to associate QSOs with galaxies of known redshift and to show the redshifts are equal (Gunn 1971), and hence cosmological, or unequal (Burbidge et al. 1971), and hence intrinsic. The data are again confused. The absence of line variation in QSOs argues for a large scale and hence a large (cosmological) distance. Inadequate observational data are available for a firm conclusion to be reached.

3.2 Emission Line Models

A considerable effort has been put into calculations of emission line intensities on the "modified planetary nebula" model (Bahcall and Kozlovsky (1969), Williams (1967), Davidson (1972), McAlpine (1972), Chan and Burbidge (1975)). A uniform distribution of matter or clouds is supposed to surround a central source with a power law spectrum. The models have been plagued by the absence of adequate observational data for comparison purposes and have been rather over-dependent on results for 3C 273 in which there are no forbidden emission lines. Densities are thus normally deduced to be high ($\gtrsim 10^6$ cm^{-3}) and, in order to maintain the "observed" ionization, the material has to be placed close to the continuum source (∿ 10^{19} cm). To avoid problems with line intensities and electron scattering optical depth, clumsy models (filling factors ∿ 10^{-4}) are preferred. These models usually make no prediction of line profiles which, as we have seen above, can be very different even in the same QSO.

The difference in [OIII] and Balmer line profiles in low
redshift QSOs (Baldwin 1975) in fact suggests that there must be
at least two regions from which emission lines arise. Furthermore,
the presence of [OII] indicates a low density and hence, an emitting
region which is more distant than that discussed above. We shall
return to this topic later in discussing both the absorption line
data and outflow models. We should, however, note that there is
no difficulty, in photoionization models, in producing approximately
correct emission line strengths with normal composition material.
The wide range of ionization follows directly from the power law
spectrum (Williams 1967) and evidence for truncation of this
spectrum, at least in some QSOs, may be adduced from the absence
of HeII $\lambda4686$ and/or NV $\lambda1240$ in many cases.

3.3 Absorption Line Data

Apart from discussions of the reality of the many absorption
line systems in high z QSOs, comparatively little has been achieved
in interpreting the data. This is partly because the lines are
frequently unresolved and saturation effects make interpretation
difficult. Nonetheless, it appears that (i) a wide range of ionization is usually present, (ii) the relative abundances of the metals
are not abnormal (although there may be some evidence of an overall
deficiency with respect to hydrogen), (iii) lines from excited levels,
even fine structure levels of the ground state, are very rarely seen
and (iv) most absorption systems are at a blueshift relative to the
emission system.

If, as seems likely, most of the absorption is intrinsic to
the QSO, it presumably must arise in outward flowing material.
The origin of this outflow is still unclear but the most plausible
suggestion so far is that it is due to radiation pressure, as in
the atmospheres of early type stars (Lucy and Solomon 1970). As
pointed out by Scargle et al. (1970) this could lead to certain
preferred relative velocities (or values of R) – the line locking
process – and there has therefore been some effort to search for
these (Burbidge and Burbidge 1975). The most obvious possible
cases are the resonance doublet splittings at 500–1000 km/sec as
was observed, for example, in CIV in 3C 191 (Williams et al. 1975)
and MgII in 1331+170 (Carswell et al. 1975). Note that these
splittings can occur in absorption systems with redshifts both
close to and well-shifted from the emission value. New high resolution data have shown splittings of order 100–200 km/sec, for
example in 0237–23 (Boksenberg and Sargent 1970), with perhaps
a preferred value at 140 km/sec. Rees (1975) has attributed this
to Ly-α locking to the nearby HeII transition. Finally, there is
some suggestion that there may be lockings at high values of R, in
particular at R = 1.11 and 1.32 corresponding to Ly-α and Ly-β
respectively locking at the Lyman continuum (e.g. PHL 957 and 1331+
+170, Coleman et al. 1976). The R = 1.11 ratio appears to be present in many QSOs (Burbidge and Burbidge 1975) and is certainly the

easiest to test as well as being plausible physically if HeII
rather than hydrogen is involved.

The absence of lines from excited fine structure levels of
the ground state enables one to set fairly stringent limits on
conditions in the absorbing region (Bahcall and Woolf 1968). Such
levels may be populated by collisions, UV flourescence or direct
IR excitation of which the first two are most likely to occur in
QSOs. The low population of such excited levels in QSOs set
limits on the density (n << 10^3 cm^{-3}) or to the distance of the
cloud from the continuum source R >> 10^{22} cm. The ionization state
provides another relation between density and distance from the
continuum source. The only QSO in which excited fine structure
level population is clearly established is 3C 191 (Burbidge, Lynds
and Burbidge 1966). In this case, arguments along the above lines
indicate that $n_e \sim 10^3$ cm^{-3} and R $\sim 10^{22}$ cm (Williams et al. 1975).
Furthermore, if the absorbing clouds cover a reasonable fraction
of the source, it can be shown that the absorbing material must
contribute substantially to the emission line flux (Williams et al.
1975). The absorption line data may thus also contain hints that
part of the observed emission spectrum comes from clouds which are
comparatively far from the continuum source.

3.4 Flow Models

A few remarks on the radiation pressure mechanism are probably
in order. The outward acceleration may be a consequence of resonance line or continuum absorption or indeed electron scattering,
although the latter is usually small. For the resonance lines,
the acceleration is

$$g_{R,L} = \frac{\pi c^2}{m_e c^2} \sum_{a,i,\ell} F_\nu \frac{n(a,i)}{\rho} f_\ell \qquad (3.1)$$

where F_ν is the flux, f the oscillator strength, n(a,i) the number
density of element a in ionization state i, subscript ℓ denotes
the line in question and other symbols have their usual meaning.
For the continuum

$$g_{R,C} \simeq \sum_{a,i} \alpha_\nu \frac{n(a,i+1)}{\rho} \frac{h\nu_C}{c} n_e \qquad (3.2)$$

where α_ν is the recombination coefficient and ν_C is the frequency
at the ionization edge (Mushotzky et al. 1972, Kippenhahn et al.
1974). Since the integrated oscillator strength in the continuum
is close to that in a resonance line, there is roughly one ionization
per resonance scattering if the spectrum is of power law form with
$\alpha \sim 1$. The acceleration from lines and continuum must be comparable
and be determined by the recombination rate. Unless the densities
are high ($n_e \sim 10^9$ cm^{-3}) this is much longer than a scattering time.
Ionization thus reduces the efficiency of the process. If, on

the other hand, there is a high energy cut-off in the spectrum so that an upper limit exists to the ionization, the resonance lines can act to their full efficiency provided they do not become optically thick.

The question of whether acceleration takes place in clouds or in smooth flows is not yet resolved although it does seem that the material must eventually break-up into clouds. The steady state situation is better defined mathematically and has therefore received more attention recently (Kippenhahn et al. 1975, Williams and Strittmatter 1976). The spectrum of PHL 5200 indicates, that at least in this case, a continuous flow may be occuring. The equation of motion for such a flow is

$$\frac{dv}{dr} = \frac{v}{v^2 - v_s^2} \left[\frac{k}{\mu m_H} \left(\frac{2T}{r} - \frac{dT}{dr} \right) - g + g_R \right]$$ (3.3)

At the sonic point it follows that $g \sim g_R$ and that if large accelerations are to occur a substantial increase in g_R is required as v increases from v_s. In most situations, however, acceleration results in a decrease in g_R so that v stays near the sonic value. Large velocities can be reached, however, if there is an "after-burner" effect (i.e. an increase in g_R) either due to cut-off in the continuum (as in stars) or due to luminosity generated outside the sonic point.

It is instructive to ask what the immediate environment of a QSO would be like if such outflow were to occur. One would expect a high density interior zone where acceleration is taking place and where the broad emission lines are formed. This would be followed by a low density high velocity zone from which comparatively little can be observed. Finally a shock should occur at large distances where the high velocity material meets the ambient gas of previously ejected material. It would be in this zone that the absorption would occur. This picture is not in conflict with any current information on QSOs but clearly needs much more investigation. It would, however, account for the apparent split in line formation regions between regions of order parsecs (broad emission) and those at several kiloparsecs (sharp emission, absorption).

Research on QSOs at Steward Observatory is supported by the NSF.

REFERENCES

Bahcall, J.N., 1971 Astron. J. 76, 283.
Bahcall, J.N. and Kozlovsky, B., 1969, Ap. J. 155, 1077.
Bahcall, J.N. and Wolf, R.A., 1968, Ap. J. 152, 701.
Baldwin, J.A., 1975, Ap. J. 20, 26.
Boksenberg, A.E. and Sargent, W.L.W., 1975, Ap. J. 198, 31.
Burbidge, E.M. and Burbidge, G.R., 1975, Ap. J. 202.
Burbidge, E.M., Burbidge, G.R., Solomon, P.M., and Strittmatter, P.A., 1971, Ap. J. 170, 233.

Burbidge, E.M., Lynds, C.R., and Burbidge, G.R., 1966, Ap. J. 144, 447.

Burbidge, E.M., Lynds, C.R., and Stockton, A.N., 1968, Ap. J. 152, 1077.

Carswell, R.F., Hilliard, R.L., Strittmatter, P.A., Taylor, D.J. and Weymann, R.J., 1975, Ap. J. 196, 351.

Carswell, R.F., Strittmatter, P.A., Williams, R.E., Kinman, T.D. and Serkowski, K., 1974, Ap. J. Letters 190, L101.

Chan, Y.W.T. and Burbidge, E.M., 1975, Ap. J. 167, 213.

Coleman, G., Carswell, R.F., Strittmatter, P.A., Williams, R.E., Baldwin, J., Robinson, L.B. and Wampler, E.J., 1976, Ap. J. (in press).

Davidson, K., 1972, Ap. J. 171, 213.

Disney, M.J., Peterson, B.A. and Rodgers, A.W., 1974, Ap. J. Letters 194, L79.

Gunn, J.E., 1971, Ap. J. Letters 164, 413.

Hazard, C., Mackey, M.B. and Shimmin, A.J., 1963, Nature 197, 1037.

Hoyle, F., Burbidge, G.R. and Sargent, W.L.W., 1966, Nature 209, 751.

Jones, T.W., O'Dell, S.L. and Stein, W.A., 1974, Ap. J. 188, 253.

Kinman, T.D., 1975, Variable Stars and Stellar Evolution (ed. Sherwood and Plant) 573.

Kippenhahn, R., Mestel, L. and Perry, J.J., 1975, Astron.&Astrophys. 44, 123.

Kippenhahn, R., Perry, J.J. and Röser, H.J., 1974, Astron.&Astrophys. 34, 211.

Kristian, J., 1973, Ap. J. Letters 179, L61.

Lowrance, J.O., Morton, D.C., Zucchino, P., Oke, J.B. and Schmidt, M., 1972, Ap. J. 171, 233.

Lucy, L.B. and Solomon, P.M., 1970, Ap. J. 159, 879.

Lynds, C.R., 1971, Ap. J. 147, 396.

MacAlpine, G.M., 1972, Ap. J. 175, 1.

Mathews, T.A. and Sandage, A.R., 1963, Ap. J. 138, 30.

Morton, D.C. and Morton, W.A., 1972, Ap. J. 174, 237.

Mushotzky, R.F., Solomon, P.M. and Strittmatter, P.A., 1972, Ap. J. 176, 7.

Oke, J.B., Neugebauer, G. and Becklin, E.E., 1970, Ap. J. 159, 341.

Rees, M.J., 1975, M.N.R.A.S. 171, 1P.

Savage, B.D. and Jenkins, E.B., 1972, Ap. J. 1972, 491.

Scargle, J.D., Caroff, L.J. and Noerdlinger, P.D., 1970, Ap. J. Letters 161, L115.

Schmidt, M., 1963, Nature 197, 1060.

Wampler, E.J., Robinson, L.B., Burbidge, G.M. and Baldwin, J.A., 1975, Ap. J. Letters 198, L49.

Weymann, R.J. and Williams, R.E., 1976, preprint.

Williams, R.E., 1967, Ap. J. 147, 556.

Williams, R.E. and Strittmatter, P.A., 1976, in preparation.

Williams, R.E., Strittmatter, P.A., Carswell, R.F. and Craine, E.R., 1975, Ap. J. 202.

RADIO SOURCES AND CLUSTERS OF GALAXIES

Wallace L. W. Sargent

Hale Observatories
California Institute of Technology
Carnegie Institution of Washington

1. INTRODUCTION

Almost all theories of extended extragalactic radio sources
involve the idea of confinement by a tenuous external medium.
Generally it is hard to pin down the existence and properties
of such a medium. However, in clusters of galaxies there are
various lines of evidence, notably derived from X-ray observations, that point to the existence of gas which would
exert a considerable influence on the evolution of radio sources.
In these lectures we shall summarize some relevant optical,
radio and X-ray properties of clusters. Then we discuss the
evidence for various forms of gas in clusters. Finally, we
discuss in detail observations of the Perseus cluster which is
particularly well studied.

2. CLUSTERS OF GALAXIES

2.1 Optical properties of clusters

The two principal compilations of clusters of galaxies are by
Abell (1958) and the six-volume "Catalogue of Galaxies and
Clusters of Galaxies" (CGCG) by Zwicky and his collaborators.
The CGCG contains 9700 clusters. All of them are relatively
rich, having at least 50 galaxies in the magnitude range m_{max} to
$m_{max} + 3$ where m_{max} is the apparent magnitude of the brightest
galaxy. Unfortunately, the populations estimated by Zwicky's
prescription are distance dependent so that the catalogue cannot

G. Setti (ed.), The Physics of Non-Thermal Radio Sources, 67–81. All Rights Reserved.
Copyright © 1976 by D. Reidel Publishing Company, Dordrecht-Holland.

be used for some statistical purposes. On the other hand, Abell
paid strict attention to the problem of statistical completeness
in compiling his list of 2712 clusters
     The Abell clusters are all very rich, having more than 30
galaxies between m_3 and $m_3 + 2$ magn, where m_3 is the apparent
magnitude of the third brightest galaxy. The clusters are sub-
divided into "richness classes" R running from 0 to 5 according
to the following Table:

Table I

Abell Cluster Richness Classes

R	0	1	2	3	4	5
Population[*]	30-49	50-79	80-129	130-199	200-299	>300
No. of clusters		1224	389	68	6	1

[*] Number of galaxies with apparent magn. between m_3 and $m_3 + 2$.

Abell also divided the clusters into "distance groups" D according to the apparent magnitude m_{10} of the tenth brightest
galaxies. The following Table shows the distribution after
distance group together with the mean estimated redshift and
mean m_{10} for each group:

Table II

Abell Cluster Distance Groups

D	1	2	3	4	5	6
No. of clusters	9	2	33	60	657	921
$\langle z \rangle$	0.027	0.038	0.067	0.090	0.140	0.180
$\langle m_{10} \rangle$	13.26	14.40	15.36	15.96	17.02	17.64

     Abell showed that the galaxies in rich clusters obey a
characteristic distribution over luminosity. Brighter than a
characteristic luminosity m_*, the number of galaxies brighter
than apparent magnitude m is of the form $N(<m) \sim m^{0.8}$. Fainter
than m_* the relation is $N(<m) \sim m^{0.2}$. This integral luminosity

function has a sharp "knee" at magnitude m^* which is about 2.5
magn. fainter than the brightest galaxy in a cluster. There are
several analytic representations of the typical cluster luminosity function, which is fairly flat for the faint galaxies and
steeply falling for the bright ones. A particularly useful
representation is one recently found by Schecter (1974). Let
$\Phi(L)$ dL be the number of galaxies per unit volume with luminosities in B light between L and L+dL. Schecter's luminosity
function has the form

$$\Phi(L/L_*)d(L/L_*) = \Phi^* (L/L_*)^{-\alpha} e^{-L/L_*} d(L/L_*). \qquad (2.1)$$

For rich clusters the value of α is 5/4 while $L_* =$
3.4×10^{10} $L_\odot$. The equivalent absolute magnitude M_* is -20.85.
(Here, and elsewhere in these notes we use a Hubble constant of
$H_o = 50$ km sec^{-1} Mpc^{-1}.)

We have seen that the Abell clusters are all very rich.
There has been little systematic work on poorer clusters and
loose groups of galaxies. In fact, systematic searches for less
rich clusters have only been made for groupings within 40 Mpc
of the local group. De Vaucouleur's (1976) list of nearby groups
has existed in preprint form for several years and has been
analyzed by several workers (see e.g., Turner and Sargent (1974)).
More recently, Gott and Turner (1976) have analyzed the distribution of galaxies in the Zwicky catalogues and have produced a list of groups, similar to de Vaucouleur's but based on
more impersonal criteria. They found that, statistically, the
loose groups of small numbers of galaxies obey a luminosity
function like Schecter's with $\alpha \sim 1.0$. Moreover, Gott and
Turner have derived a multiplicity function--the ratio of the
space density of single galaxies to binaries to triples, etc.,
all the way up to the rich Abell clusters. As an example of
their findings they state that about 10 percent of the light of
the Universe comes from rich clusters of the Abell type with
$L > 300$ L_*.

2.2 Types of clusters

In the absence of a physical theory of the formation and evolution of clusters of galaxies, the classification of these
objects is a purely empirical art. Two important classification
schemes have been proposed by Bautz and Morgan (1970) and by
Rood and Sastry (1971). For complete descriptions of these
schemes, the reader is referred to the papers cited above.
However, in outline the Bautz-Morgan scheme categorizes clusters
according to the degree to which they contain a dominant supergiant elliptical (cD) galaxies. Type I clusters contain a
dominant cD (e.g., Abell 2199, Abell 2029), Type II clusters

contain brightest galaxies intermediate between a cD and a
Virgo giant elliptical (e.g., Coma, Abell 2197) while Type III
clusters contain no dominant galaxies (e.g., Virgo, Hercules,
Coronae Borealis).

Rood and Sastry have a more complex morphological scheme
in which the presence or absence of a cD galaxy is only one of
the criteria. Thus, their cD clusters resemble the Bautz-
Morgan Type I, while their I (Irregular) clusters resemble
Type III. In between Rood and Sastry define clusters of Type B
(with two dominant central galaxies), L (with a line of bright
galaxies), C (core-halo clusters with several of the very
brightest galaxies in the core) and F (several of the brightest
galaxies in a flattened configuration).

Numerically, the clusters in Abell's distance groups 0 to
3 divide up on the Rood-Sastry scheme according to the following
Table:

Table III

The Rood and Sastry Types

cD	B	L	C	F	I
23	10	10	15	20	32

2.3 Types of galaxies

There is some relationship between the appearance of a cluster
and the proportions of galaxies of different types. This has
been brought out particularly by Oemler (1974) who has divided
clusters into three categories similar to Bautz and Morgan and
who finds the following proportions according to Type:

Table IV

Proportions of Galaxies of Different Types

	Ratio		
	E:	SO:	Spiral
cD clusters	3	4	2
Spiral poor	1	2	1
Spiral rich	1	2	3

Oemler has also argued that the sequence

| Irregular cluster with no dominant central galaxy | → | Centrally condensed cluster with dominant bright central galaxy |

is one of increasing dynamical evolution. Oemler's conclusion is based on the degree to which the bright and faint galaxies are segregated at the cluster center.

Oemler derived parameters such as the total mass M, the gravitational radius R_G and the total luminosity L for eight clusters. His values are shown in the following table:

Table V

Physical Parameters of Some Rich Clusters

Cluster (Abell No. or Name)	L $(10^{12}L_e)$	R_G (Mpc)	σ (km/sec)	M/L
194	1.8	2.6	782	203 ± 47
400	3.5	3.3	359	28 ± 15
1314	3.1	2.8	928	179 ± 86
Coma	9.4	3.9	1430	195 ± 38
Hercules	3.6	4.0	749	143 ± 50
2197	4.2	4.1	490	54 ± 25
2199	5.7	3.7	1092	173 ± 49
2670	5.5	3.7	1525	360 ± 125

The gravitational radius is defined by

$$\frac{1}{R_G} = \frac{1}{L^2} \int_0^\infty \frac{L(r)dL(r)}{r} , \qquad (2.2)$$

where L(r) is the light omitted by the cluster between radius r and r+dr.

3. RADIO SOURCES AND CLUSTERS

Early studies by e.g., Mills (1960), van den Bergh (1961) and Pilkington (1964) showed a significant correlation between radio sources and Abell clusters. The most recent observations, by Owen (1974a) showed that of 503 Abell clusters with distance class < 4, 259 had continuum radio sources with a flux bigger than 0.1 Jky at 21 cm.

3.1 Correlation of optical and radio properties

Owen (1975) found no significant correlation between radio
emission and richness class for the Abell clusters. However,
Owen's survey was confined to clusters in distance groups 4 or
less: the richest Abell clusters are practically all in
distance groups 5 and 6.

Matthews, Morgan and Schmidt (1964) showed that about half
of the strong 3C radio sources (those with radio luminosities
$L_r > 10^{41}$ ergs sec^{-1}) are associated with cD galaxies in
clusters. Moreover, 7 out of the 10 cD clusters known to
Mathews et al. gave radio emission; in 5 cases the emission was
definitely associated with the cD galaxy itself.

Owen (1975) examined the correlation between radio
emission and cluster type, using both the Rood-Sastry types and
the Bautz-Morgan types. He concluded that clusters of all types
contain a significant proportion of detectable radio emission
but that clusters containing cD galaxies are more likely to be
detected than those without them.

3.2 Radio properties of X-ray clusters

Kellogg et al. (1973) listed 21 X-ray clusters; mostly found in
the UHURU survey. A few more have been identified since then.
On the basis of the current evidence it is likely that the
X-rays are produced by a hot, ionized, diffuse gas, although a
contribution from non-thermal radiation also seems likely in
some cases. Owen (1974b) has examined the correlation between
the radio, X-ray and optical properties of the ten clusters
then known which have X-ray luminosities $L_x > 10^{44}$ ergs/sec.
Owen's findings may be summarized as follows:

a) Nine out of 10 X-ray clusters are also significant radio
sources with a luminosity $z^2 S_{1400} > 5 \times 10^{-4}$ (c/H$_0$ = 1, q$_0$ = 1).
Only 30 to 40 percent of all Abell clusters are detected at this
level of radio flux.

b) Eight out of the 9 detected clusters have large radio
structures ($\gtrsim$ 5 arcminutes).

c) X-ray sources occur most frequently in clusters with a central
dominant galaxy or galaxies. The correlation with cluster type is
much more pronounced than that found at radio frequencies.

3.3 Structures of strong radio sources in clusters

Riley (1975) has investigated the **structures** of the 25 4C radio
sources identified with Abell clusters between declinations
+20° and +40°. Riley's main conclusions were as follows:

a) When there is one strong radio source it is always

associated with one of the <u>brightest</u> galaxies in the cluster.
In the case of a dominant brightest galaxy (7 examples), the
radio source is always associated with that galaxy. In clusters
with no dominant galaxy (5 examples), the radio source is
associated with <u>one</u> of the brightest galaxies.

     b) Of sources identified with a cluster galaxy, three
were unresolved. The rest may have small scale structure
(<10 arcseconds), but they also have large scale structure
50-100's of kpc in size.

     c) Four of the clusters appear to contain more than one
radio source definitely identified with a galaxy. In all cases
these radio sources are associated with the brightest galaxy
and one or more of the other bright galaxies.

3.4 Angular extent of radio sources inside and outside of
    clusters

Extragalactic radio sources are frequently found to be double
with the two components on either side of a central galaxy. It
is generally presumed that the components have been ejected from
their associated galaxy. It is still in dispute as to whether
there is any difference between the double radio sources inside
and outside clusters. De Young (1972) found that in a list of
34 sources with no obvious selection effects, the mean component separation for sources outside clusters is about twice
that for sources inside. He also found that the mean ratio of
the component size to separation for sources outside was only
60 percent of that for sources inside clusters. On the other
hand, Hooley (1974) contradicted De Young's result. Hooley took
a complete sample of 199 3CR sources with $|b| > 10$ deg. and with
$S \geq 10$. He then selected 34 radio galaxies with $m_V < 17.5$ and
with a linear size of source greater than 20 kpc. Only 5 of the
34 sources had <u>not</u> been reported to be associated with a
cluster or grouping of galaxies. Of the 16 "simple" doubles in
the sample, 4 were isolated, 4 were in Abell clusters, and
eight were in smaller clusters. From the small sample, Hooley
was not able to detect any significant difference in the
structure of sources associated with the presence or lack of
presence of a cluster.

4. GAS IN CLUSTERS OF GALAXIES

It has been known for some time that there is a different proportion of emission line galaxies in clusters and in the field.
Thus, Humason, Mayall and Sandage (1956) reported that 18 percent of E galaxies, mostly in the field, show $\lambda 3727$ [O II]
emission. Later Osterbrock (1960) noted that of 25 E galaxies
in 4 dense clusters, there was not one with $\lambda 3727$ emission.

More recently Gunn and Sargent (unpublished) have found that
out of about 150 E and SO galaxies in the Coma and Perseus
clusters, only one (NGC 4874 in Coma) has even weak $\lambda 3727$
emission.

This may be interpreted as direct evidence that the gas in
cluster E and SO galaxies is swept out by the galaxies' passage
through a tenuous intracluster medium. This conclusion is also
suggested by recent studies of the radial distribution of galaxy
types in the X-ray clusters. We saw already that the X-ray
clusters tend to be regular, centrally condensed and with a
dominant cD galaxy. Such clusters have a low proportion of
spirals as Oemler's data (Table IV) shows, and the spirals that
do exist are concentrated to the outside of the cluster. Now
it has been proposed that SO galaxies are spirals from which gas
was removed early on in their evolution either by sweeping or by
collisions with other spirals. Following this notion Melnick
and Sargent (to be published) have investigated the radial
distribution of galaxy types (E, SO and Sp) in several of the
X-ray clusters. In all cases the ratio Sp/SO rises with increasing radius and then reaches a constant level at a radius
of about 0.6 Mpc. The overall percentage of SO galaxies varies
from cluster to cluster, but it correlates with the X-ray
luminosity. The most luminous X-ray clusters have the highest
proportion of SO galaxies.

These facts again seem consistent with the general idea
that clusters contain a tenuous hot gas which sweeps out the
interstellar gas in spiral galaxies as their orbits take them
near the center of the cluster. The SO galaxies which one
supposes to be produced by this process should show a range of
integrated color, depending on how long has elapsed since their
star formation ceased. For this reason, a study of the colors
of SO and spiral galaxies in rich clusters would be of great
interest. In particular, the sweeping out hypothesis for the
formation of SO's would predict that the ones in the center
should be on average redder than those on the outside of the
cluster.

4.1 Neutral hydrogen in clusters

There have been several searches for diffuse neutral hydrogen
emission in large clusters of galaxies using the 21 cm line.
Only upper limits have been reported. For example, De Young
and Roberts (1974) made observations of the Coma, Cancer,
Pegasus I and Perseus clusters with a 22 arcminute beam and with
a receiver having a bandwidth of corresponding to 1700 km/sec
in velocity. Upper limits to the peak antenna temperature of
about 0.02 deg. were found for each object. This corresponds to
an upper limit to the amount of optically thin neutral hydrogen
of between 2×10^{10} $M_\odot$ and 4×10^{11} $M_\odot$, depending on the cluster
and on the adopted model. For the Coma cluster the upper limit

on the neutral hydrogen density is $n_H \leq 10^{-4}$ cm^{-3}. Neutral hydrogen has not been detected in individual E galaxies. The sums of the expected upper limits for the 10^2 to 10^3 galaxies in the beam would be 10^8 M$_\odot$, much less than the actual upper limit observed for the whole cluster.

Smart (1973) proposed that large clusters might contain neutral hydrogen in the form of **optically thick** clouds which have the same radial distribution as the galaxies. Smart supposed the clouds to be disks of radius R_c and thickness of $R_c x$ ($x \ll 1$). He showed that the 21 cm emissivity is

$$E_1 = 2\pi R_c^2 (1 + x) A_{21} E_{21} n_H N_b \ell , \qquad (4.1)$$

where N_b is the number of clouds in the beam and ℓ is the depth in a cloud at which the 21 cm optical depth reaches unity.

The observed emission is

$$E_2 = \frac{2kT_B}{\lambda^2} \sigma_\nu (2\pi)^{1/2} \Omega_s 4\pi D^2 \text{ ergs/sec} , \qquad (4.2)$$

where T_b is the peak antenna temperature, σ_ν is the frequency spread in the line and Ω_s is the angular size of the source.

De Young and Roberts showed that their data would admit the existence of optically thick clouds having a total mass of about 10^{13} M$_\odot$. The hypothetical clouds would have a diameter of about 20 kpc, $x \sim 0.1$, and a mass $M_c \sim 10^{12}$ M$_\odot$. Such objects may be implausible on theoretical grounds and, in any case, would not affect the dynamical evolution of extended radio sources in clusters.

There has been little observational work on intergalactic neutral hydrogen in poor clusters of galaxies. Recently, however, Mathewson, Cleary and Murray (1975) discovered large H I clouds about 100 kpc away from NGC 55 and NGC 300 in the Sculptor group. This is one of the nearest groups of galaxies. The clouds have a maximum column density of 1.2×10^{20} H atoms cm^{-2}, a volume density of about $n_H = 8 \times 10^{-4}$ cm^{-3} and internal velocity dispersions of only 35 km/sec. The largest clouds around the two Sculptor galaxies have masses of about 10^9 M$_\odot$, comparable to the total mass of neutral hydrogen in the disk of a typical spiral galaxy.

It is likely that such large aggregates of neutral hydrogen are confined to the poorer clusters of galaxies. In the large clusters the intergalactic gas is probably all heated and ionized by the combined effects of infall and the virial motions of the galaxies.

4.2 Molecular hydrogen in clusters

There has been no direct detection of extragalactic molecular hydrogen. Noonan (1972) has argued that any diffuse molecular

hydrogen in large clusters would be photodissociated by ultraviolet radiation from the galaxies. According to Noonan's
estimates the flux of UV radiation is 10^4–10^5 times too large
for H_2 molecules to exist either inside or outside clusters in
intergalactic space.

4.3   Ionized hydrogen in clusters

Several observational approaches have been used in attempts to
detect ionized gas in clusters of galaxies:
   a)   X-ray emission from a hot gas 10^7–10^8 deg.
   b)   Lyα recombination radiation from gas 10^4–10^6 deg.
   c)   Hβ recombination radiation, also from gas 10^4–10^6 deg.
   d)   Radio continuum emission, gas hotter than 10^4 deg.
   c)   Inverse Compton cooling of the microwave background
radiation in the direction of the Coma cluster.
       All these methods give upper limits to the hot gas density
except for the X-ray observations. Also essentially all of the
existing literature discusses the observations in terms of the
amount of ionized gas required to bind the clusters gravitationally. In order to bind a cluster like Coma, one requires
about 2 x 10^{15} $M_\odot$ of ionized hydrogen contained in a radius of
about 4 Mpc. This corresponds to a mean electron density
$\langle n_e \rangle \sim 10^{-3}$ cm^{-3}. In fact, there seems to be at most 10^{-4} cm^{-3}.
This is insufficient to bind the cluster but is large enough to
have a significant effect on the "radio trail" sources, on the
general confinement of extended radio components and to sweep
out gas from spiral galaxies moving at around 1000 km/sec.
       The X-ray measurements indicate that a very hot gas, in
which the thermal velocity is comparable to the virial velocities
of the galaxies, is present in the great clusters. In order to
set limits on cooler ionized material it is necessary to combine
the results obtained from the types of measurements listed
earlier. As an example, one may cite the work of Crane and
Tyson (1975). They obtained an upper limit for the Hβ
emissivity from four clusters, Abell 779, Abell 1367, Coma and
Abell 2199. (Presumably, Hβ is measured rather than Hα because
it falls at a wavelength where the night sky contains little
emission and where blue sensitive detectors, of high quantum
efficiency, may be employed.) They find for all the 4 clusters
a 2σ upper limit for the Hβ flux of 2 x 10^{-18} ergs sec^{-1} cm^{-2}
$arcsec^{-2}$. A similar upper limit is found for the optical
continuum near Hβ. Crane and Tyson combine these measurements
with the radio continuum observations of Davidsen and Welch
(1974) and with the X-ray observations of Kellogg et al. (1973).
They find that in the cases of Abell 1367 and Abell 2199 it is
not quite possible to exclude the virial mass of ionized gas
($\sim 10^{15}$ $M_\odot$) at any temperature. However, if the gas has a
thermal velocity equal to the velocity dispersion of the

galaxies, as expected, then the gas must be clumpy if it has
the virial mass. This is an unlikely state of affairs. In the
case of the Coma cluster the soft X-ray flux limit by
Gorenstein et al. (1973) may be combined with the Hβ limit to
show that the mass of ionized gas is less than 24 percent of the
virial mass even if it is not clumped. If it is clumped then
since the emissivity goes as n_e^2 a lower mass would be
indicated.

The Perseus cluster, which is of particular interest in
radio astronomy because it contains the radio galaxy NGC 1275
(Per A) as well as weaker radio galaxies, is a more difficult
case. Observations of soft X-rays and of the space distribution of X-rays by Wolff et al. (1974, 1975), by Fabian
et al. (1974) and by Davidsen et al. (1975) indicate that there
is extended emission from the cluster together with harder
X-rays from NGC 1275 itself. Analysis of the available data
by Davidsen et al. show that the upper limit to the mass of hot
gas in the cluster is 6×10^{13} $M_\odot$ or 3 percent of the virial
mass. Again, if the gas is clumpy, an even lower mass is
required to produce the observed X-ray emission.

5. REMARKS ON THE PERSEUS CLUSTER

The Perseus cluster of galaxies (Abell 426) contains a number
of radio sources, including the Seyfert galaxy NGC 1275 (Per A).
This cluster has several unusual and inexplicable properties.
Perseus is symmetrical and centrally condensed in the large and
NGC 1275, the brightest galaxy, is at the center according to
N. Bahcall (1974). However, the brighter galaxies near the
center of the cluster are obviously distributed along an axis
with NGC 1275 at one end. At the other end of this axis, which
is about 1 Mpc long, is the radio tail galaxy IC 310.
Bahcall's counts show that this elongation is also present in
the distribution of faint galaxies; the axial ratio of the
distribution is 2:1. Chincarini and Rood (1971) discussed
measurements of 49 redshifts for galaxies in the Perseus
cluster. They showed that the mean redshift of the cluster is
5460 km sec^{-1}. This is almost identical to the redshifts of
NGC 1275 and IC 310 which are 5291 and 5328 km sec^{-1},
respectively. However, NGC 1265, a second radio tail galaxy
in the system, has the much larger redshift of 7656 km sec^{-1}.
Chincarini and Rood also found that the cluster has a very
large velocity dispersion; they obtained a radial component of
1420 ± 140 km sec^{-1}. They also pointed out that the large
velocity dispersion is hard to reconcile with the asymmetric
form of the cluster center. Moreover, the large velocity of
NGC 1265 implies a very large mass for the cluster, about
2×10^{15} $M_\odot$, if it is gravitationally bound.

An inspection of Chincarini and Rood's data reveals that
there is a possibility that the velocity distribution in the
cluster has two peaks. If true this would imply that Perseus
might, in fact, be two clusters accidentally superimposed on
the plane of the sky but dynamically separate. In order to
investigate this possibility the writer obtained 30 new redshifts, paying particular attention to galaxies in the axis
or chain and to galaxies on the side of NGC 1275 away from the
chain. Table VI gives the number of galaxies in each 500
km sec^{-1} interval of redshift for the new data combined with
that of Chincarini and Rood.

Table VI

Redshifts for Perseus Cluster Galaxies

Redshift Range km sec^{-1}	No.	Redshift Range km sec^{-1}	No.
2000 – 2500	2	5500 – 6000	6
2500 – 3000	0	6000 – 6500	11
3000 – 3500	4	6500 – 7000	7
3500 – 4000	3	7000 – 7500	6
4000 – 4500	10	7500 – 8000	3
4500 – 5000	9	8000 – 8500	1
5000 – 5500	16	8500 – 9000	1

These improved data do not lend support to the idea that
Perseus is in reality two clusters. Moreover, a detailed
inspection shows that there is no evidence that the axis of
galaxies is dynamically distinct: nor is it rotating. The
redshift of NGC 1275 is on the edge of the distribution; there
is thus no reason to suppose that it is not a cluster member.

In Table VII we show the radial variation of the
measured dispersion in radial velocity. (One arcminute is
roughly 30 kpc at the distance of the cluster.)
The overall velocity dispersion, $\sigma_r = 1246$ km sec^{-1}, is a
little smaller than Chincarini and Rood's value. It is not
significantly different from the value for the "chain" or "axis"
galaxies alone, $\sigma_r = 1101$ km sec^{-1}. Generally there is a
tendency for the velocity dispersion to decrease outwards.
This behavior has also been found in the case of the Coma
cluster and is expected on theoretical grounds.

Table VII

Velocity Dispersion σ_r as a Function of Radius R

R arcmins.	No. of galaxies	σ_r km sec^{-1}
≤5	10	843
5.1 - 10	16	1602
10.1 - 15	14	1394
15.1 - 25	21	1203
>25.1	18	1089
All	79	1246
"axis"	16	1101

The velocity dispersion for the galaxies in the very center of Perseus is considerably less than that for galaxies in the next rings outwards. This probably indicates that most of the galaxies within a few arcminutes of NGC 1275 are satellites. If this assumption is made it is possible to use the virial motions of these satellites in order to measure the mass to light ratio for NGC 1275 and its immediate environment. The result is a mass of 2.10^{13} $M_\odot$ and a mass to light ratio of about M/L = 100. Thus the cluster would need about 100 galaxies as massive as NGC 1275 in order to gravitationally bind NGC 1265. This does not seem conceivable.

5.1 NGC 1275 as a collision

We have mentioned that NGC 1275 lies at the center of the Perseus cluster and that it has the same redshift as the center of mass of the cluster. NGC 1275 is a Seyfert galaxy, but unlike most Seyfert galaxies its spectroscopic abnormality is not confined to a bright nucleus at the center. Minkowski (1957) discovered that NGC 1275 is surrounded by an extensive emitting nebulosity which, moreover, contains material with two quite distinct velocities. Photographs taken by Lynds (1970) through narrow filters in Hα light show that there is filamentary material all around NGC 1275 with a velocity close to that of the nucleus namely, 5300 km sec^{-1}. It extends out to about 140 arcseconds, or 70 kpc. In addition, there is material with a velocity close to 8000 km sec^{-1} which is less extensive and which is concentrated to the N and particularly the NW of the nucleus. Measurements by Minkowski (1957) and by Burbidge and Burbidge (1965) show that the velocity spread among the high velocity clouds is only about 200 km sec^{-1}, and that in the low velocity material is about 600 km sec^{-1}.

Minkowski interpreted NGC 1275 as a collision between two galaxies moving at a relative speed of 3000 km sec^{-1}. According to Minkowski "The system consists of a tightly wound spiral of early type and a strongly distorted late-type spiral. Inspection of blue and red exposures shows some unusual features regarding the colors and distribution of emission patches connected with the distorted spiral arms of the late-type system."

Succeeding savants have not agreed with Minkowski's interpretation. Both Burbidge and Burbidge (1965) and Lynds (1970) concluded the observed phenomena are caused by an explosion. However, De Young, Roberts and Saslaw (1973) found a sharp 21 cm absorption line at the 8000 km sec^{-1} velocity. It is hard to avoid the conclusion that the 8000 km sec^{-1} is in front of the nucleus of NGC 1275 and that this gas is falling into the center at a speed of about 3000 km sec^{-1}. In view of the cluster's large velocity dispersion, such a velocity would not be implausible for an object of galactic size moving through the center of the cluster.

If the nebulosity around NGC 1275 indeed represents a collision, it is hard to understand the precise geometry. A collision at 3000 km sec^{-1} would cause the gas in the two galaxies to merge into a hot cloud with a temperature of about 3×10^8 deg. Although there have been no studies of the physical constitution of the two clouds in NGC 1275, the reported spectra are clearly inconsistent with such a high temperature. The Perseus cluster is still a mysterious object.

REFERENCES

Abell, G. O., 1958, Ap. J. Suppl. 3, 211.
Bahcall, N. A., 1974, Ap. J., 187, 439.
Bautz, L. P., and Morgan, W. W., 1970, Ap. J. Letters, 162, L149.
Burbidge, G. R., and Burbidge, E. M., 1965, Ap. J., 142, 1351.
Crane, P., and Tyson, J. A., 1975, Ap. J. Letters, 201, L1.
Davidsen, A., and Welch, W., 1974, Ap. J. Letters, 191, L11.
Davidsen, A., Bowyer, S., and Welch, W., 1975, Ap. J., 198, 1.
de Vaucouleurs, G., 1976, in Stars and Stellar Systems, Vol. 9, ed. A. R. Sandage and J. Kristian (Chicago: University of Chicago Press).
De Young, D., 1972, Ap. J. Letters, 173, L7.
De Young, D., Roberts, M. S., and Saslaw, W. C., 1973, Ap. J., 185, 809.
De Young, D. S., and Roberts, M. S., 1974, Ap. J., 189, 1.
Fabian, A. C., Zarnecki, J. C., Culhane, J. L., Hawkins, F. J., Peacock, A., Pounds, K. A., and Parkinson, J. H., 1974, Ap. J. Letters, 189, L59.

Gorenstein, P., Bjorkholm, P., Harris, B., and Harnden, F. R., 1973, Ap. J. Letters, 183, L57.

Gott, R., and Turner, E. W., 1976, private communication.

Hooley, T., 1974, M.N.R.A.S., 166, 259.

Humason, M. L., Mayall, N. O., and Sandage, A. R., 1956, A.J., 61, 97.

Kellogg, E., Murray, S., Giacconi, R., Tannenbaum, H., and Gursky, H., 1973, Ap. J. Letters, 185, L13.

Lynds, C. R., 1970, Ap. J. Letters, 159, L151.

Mathewson, D. S., Cleary, M. N., and Murray, J. D., 1975, Ap. J. Letters, 195, L97.

Matthews, T. A., Morgan, W. W., and Schmidt, M., 1964, Ap. J., 140, 35.

Mills, B. Y., 1960, Austr. J. Phys., 13, 550.

Minkowski, R., 1957, I.A.U. Symposium No. 4, Radio Astronomy, ed. H. C. van der Hulst (Cambridge: Cambridge University Press), p. 101.

Noonan, T. A., 1972, Ap. J., 171, 209.

Oemler, A., 1974, Ph.D. Thesis, California Institute of Technology.

Osterbrock, D. E., 1960, Ap. J., 132, 325.

Owen, F. N., 1974a, A.J., 79, 427.

Owen, F.N., 1974b, Ap. J. Letters, 189, L55.

Owen, F. N., 1975, Ap. J., 195, 593.

Pilkington, J. D. H., 1964, M.N.R.A.S., 128, 103.

Riley, J., 1975, M.N.R.A.S., 170, 53.

Rood, H. J., and Chincarini, G., 1971, Ap. J., 168, 321.

Rood, H. J., and Sastry, G. N., 1971, P.A.S.P., 83, 313.

Schecter, P., 1974, Ph.D. Thesis, California Institute of Technology.

Smart, N. C., 1973, Astron. Ap., 24, 171.

Turner, E. L., and Sargent, W. L. W., 1974, Ap. J., 194, 587.

van den Bergh, S., 1961, Ap. J., 134, 970.

Wolff, R. S., Heleva, H., Kifune, T., and Weisskopf, M. C., 1975, Ap. J. Letters, 197, L99.

Wolff, R. S., Heleva, H., and Weisskopf, M. C., 1974, Ap. J. Letters, 193, L53.

THE RADIO LUMINOSITY FUNCTION OF GALAXIES

R.D. Ekers

Kapteyn Astronomical Institute, University of Groningen,
Groningen, The Netherlands

1. INTRODUCTION

In this lecture I wish briefly to outline the main methods used
to determine the radio luminosity function. I will then comment
on how the form of the luminosity function has important effects
on the properties of a sample of radio galaxies identified in a
radio catalogue. Finally I will give a few examples of the types
of problems that can best be studied using the radio luminosity
function.

2. RADIO LUMINOSITY FUNCTION FROM RADIO SOURCE CATALOGUES

The most common way in which we have obtained samples of radio
emitting galaxies is by making optical identifications of radio
sources in a catalogue. Such a catalogue is normally complete to
some limiting flux density, say S_{lim}. An identified galaxy can be
observed optically to determine its distance D, usually by measuring the redshift and using the Hubble relation. We can then
determine the monochromatic radio power $P_\nu \propto S D^2$ (ignoring cosmological corrections for the moment). The radio luminosity could
then be obtained by integrating this monochromatic power over the
radio frequency spectrum however, since this involves a somewhat
arbitary definition of the range of the radio spectrum and knowledge of the radio frequency spectral distribution, I will use
the monochromatic radio power at a given observed frequency. As
is common practice I will still call this the radio luminosity
function. Fig.1 is a sketch of the kind of number distribution of

G. Setti (ed.), The Physics of Non-Thermal Radio Sources, 83–91. All Rights Reserved.
Copyright © 1976 by D. Reidel Publishing Company, Dordrecht-Holland.

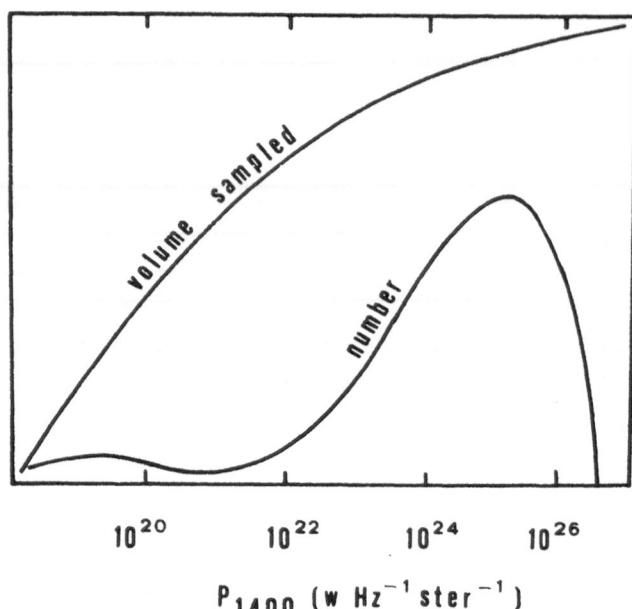

Fig. 1. Number distribution of radio galaxies found in a radio
catalogue and the effective volume sampled by the catalogue as a
function of the radio power at 1400 MHz.

radio emitting galaxies found in a complete catalogue of radio
sources (e.g. Caswell and Wills 1967, Merkelijn 1971). The main peak
at 10^{25} W Hz^{-1} ster^{-1} is caused by the objects generally referred to
as "radio galaxies". There is a secondary less well defined peak
corresponding to the "normal galaxies" at $P_\nu \sim 10^{19}$ W Hz^{-1} ster^{-1}.
This function is often referred to as the luminosity distribution
(Longair 1966). It is most important to emphasize that this is not
the luminosity function since the volume of space sampled is a function of both P_ν and S_{lim} and thus the relative number of objects in
each range of P_ν does not reflect the relative space densities of
these objects. The volume sampled for $S_{lim} = 10^{-26}$ W Hz^{-1} m^{-2} is
shown by the curved line and it can be seen that a sample such as
this is dominated by the rarer but more powerful objects that can
be seen to greater distance.

In order to convert this observed luminosity distribution to
the luminosity function it is necessary to correct for the volume
sampled. This could be done just by dividing the luminosity distribution by the average volume sampled, but the following contribution
and more accurate procedure is normally used. For each object in the
sample the maximum distance is computed, and hence maximum volume
$V_{i,max}$, for which this object would still be in the sample. This object makes a contribution of $\rho_i(P) = (V_{i,max})^{-1}$ to the luminosity
function so the final estimate of the luminosity is

$$\rho(P) = \Sigma \rho_i(P)$$

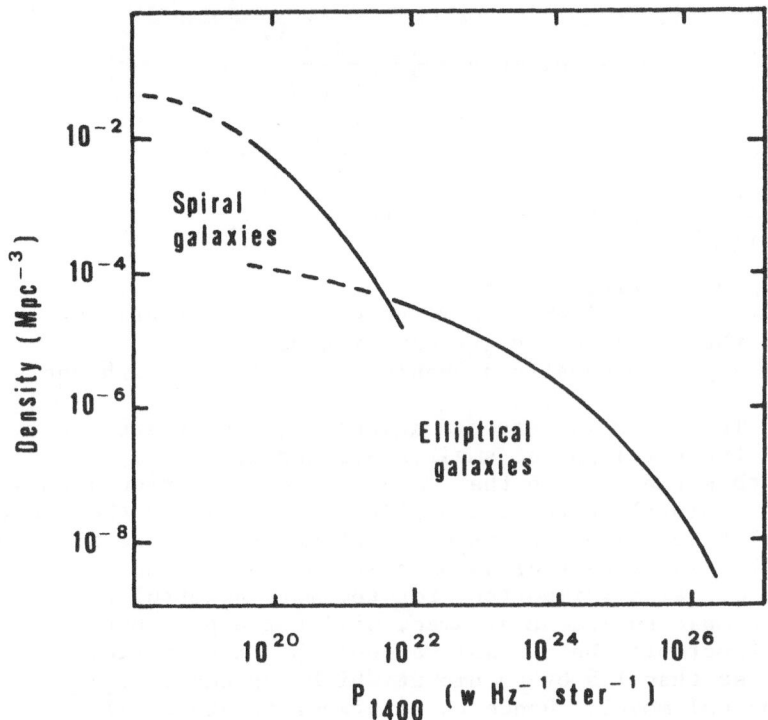

Fig. 2. Luminosity function for spiral and for elliptical galaxies.

provided the sample of objects is complete according to the criterion
used to estimate $V_{i,max}$. This method has the additional advantage
that other reasons for incompleteness, e.g. the optical magnitude
limit for complete identification, can be included by using which-
ever limit gives the smaller V_{max} (cf.Schmidt 1968). Fig. 2 shows
such a luminosity function for spiral and for elliptical galaxies.
Note that this function looks very different from the luminosity
distribution of Fig. 1. For many of the sources in a radio catalogue
V_{max} is sufficiently large that cosmological effects on the volume
sampled are important, hence the curvature of the volume sampled line
in Fig. 1. It is therefore necessary to assume a cosmological model
in order to determine the radio luminosity function, and this is particularly unfortunate if the radio luminosity function is itself required to unravel some relation (e.g. the log N – log S relation)
which is being used as a cosmological probe. A possible solution
may be to use lower luminosity galaxies as cosmological probes since
these are sufficiently numerous to determine the luminosity function
locally.
        It is interesting to look more closely at the effect of the
slope of the radio luminosity function on the observed distribution
of radio sources. In a catalogue complete to a flux density S_{lim}
we see out to a distance

$$R = \left(\frac{P}{S_{lim}}\right)^{\frac{1}{2}}, \text{ i.e. a volume } v = \frac{4}{3}\pi\left(\frac{P}{S_{lim}}\right)^{3/2}.$$

If the radio luminosity function is $\rho(P) \propto P^{-\alpha}$ then we see

$$N \propto P^{-\alpha}\left(\frac{P}{S_{lim}}\right)^{3/2} \text{ sources in the catalogue.}$$

Hence we have three possibilities:
   i) $\alpha < 1.5$. Then as P increases N increases, i.e. there are more sources in the catalogue at greater distance.
   ii) $\alpha = 1.5$. Any P is equally probable (cf Hoyle and Burbidge 1970).
   iii) $\alpha > 1.5$. Then as P increases N decreases, i.e. there are more sources in the catalogue at smaller distances.
As a result of this the peak in the luminosity distribution occurs at the point (or points) in the luminosity function where the slope changes from < 1.5 to > 1.5, and it does not necessarily correspond to any physically significant class of objects. Since the cosmological effects are always important for the more powerful radio galaxies, this simple Euclidean argument will not apply, but the nature of the effect will be the same except that the critical value of α is less than 1.5 by an amount which depends on S_{lim} and the cosmological model. Hence these arguments are still relevant for the observed value, $\alpha \sim 1.2$, at the high luminosity end.

3. RADIO LUMINOSITY FUNCTIONS FROM OPTICALLY SELECTED SAMPLES

Another method is to start with an optical sample of galaxies with known distances, e.g. a bright galaxy catalogue or a cluster of galaxies, and then measure the radio emission or upper limit to the radio emission from each member of the sample. This removes the radio power bias on the volume sampled and gives a direct estimate of the radio luminosity function. This method gives the best information on the weak end of the luminosity function which is very poorly represented in a radio flux density limited catalogue (e.g. Cameron 1971, Ekers and Ekers 1973, Jaffe and Perola 1975).
   In order to construct a luminosity function over the largest possible range, it is necessary to combine the results from these two methods.

4. FORMS OF THE RADIO LUMINOSITY FUNCTION

Some confusion is caused by the variety of ways in which the luminosity function is specified. Some of the methods used are:

a) Differential form: $\rho(P)$ dP is the density of radio galaxies with power in the range P to P + dP.

b) Differential form with logarithmic intervals: $\rho(\log P)$ d log P (or $\rho(m)$ dm where m is the magnitude) or

$$\int_P^{xp} \rho(q)\ dq$$

c) Integral form: $\rho(>P) = \int_P^{\infty} \rho(q)\ dq$

The differential forms have the advantage that each point is independent and the effect of uncertainties in the estimate is much clearer. The integral form has the advantage that the type of interval need not be specified. The differential form with logarithmic intervals is convenient since it is similar to the optical luminosity functions (which usually use magnitude inter- vals). It has the same shape as the integral form and usually gives a more nearly equal number of objects per interval.

Note that for a power law luminosity function

$$\rho(P) \propto P^{-\alpha}$$

$$\rho(\log P) \propto P^{-\alpha+1}$$

$$\text{and } \int_P^{\infty} \rho(q) \propto P^{-\alpha+1}$$

In many cases it is convenient to normalize the density $\rho(P)$ to the density of galaxies of the type being considered. We then have the function describing the probability that a given type of galaxy will be a radio source of power P dP. Such a procedure is necessary if the luminosity functions from samples with different optical compositions, e.g. cluster and non-cluster galaxies are to be combined.

5. WHY DO WE WANT THE RADIO LUMINOSITY FUNCTION?

5.1 Unbiased samples

Is it better to study the common or the rare members of a class? Because of the effect of volume sampled, radio catalogues are dominated by the very rare powerful sources such as Cygnus A while we have only poor statistics for the more common lower power sources like Centaurus A.

5.2 Effect of cluster membership

An important parameter in many radio galaxy models is the density

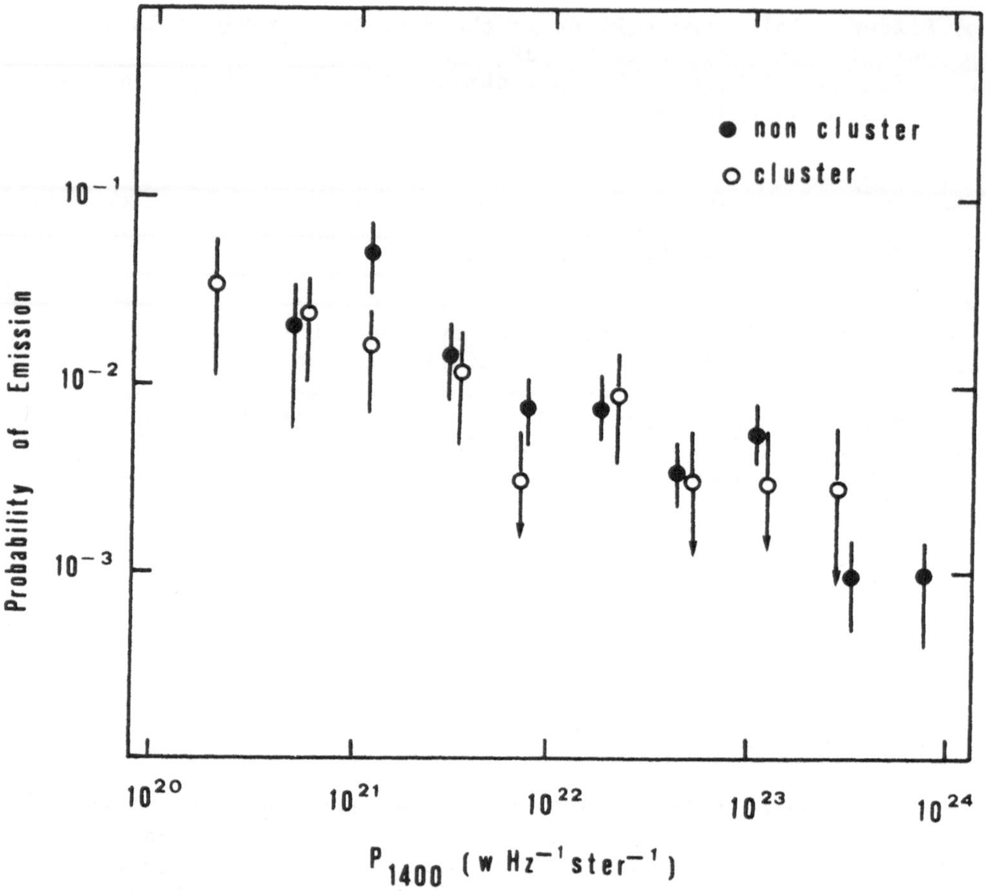

Fig. 3. Radio luminosity function for cluster and non-cluster
elliptical galaxies.

of the intergalactic medium into which they are presumed to be expanding. In order to test whether the higher density intergalactic
medium likely to exist in a rich cluster of galaxies effects their
properties, we can compare the radio luminosity functions for galaxies inside and outside of rich clusters. Such a comparison has been
made between the luminosity function in rich clusters (Jaffe and Perola 1975) and that for field clusters (Colla et al. 1975; Ekers et
al. in preparation). The result is sketched in Fig. 3 and shows no
difference greater than the statistical errors either in the shape
or value of the luminosity function.

5.3 Evolution of radio galaxies

Following the argument of Schmidt (1966), the radio luminosity function of an ensemble of galaxies may be interpreted as lifetime
of the member galaxies as a function of their radio power.

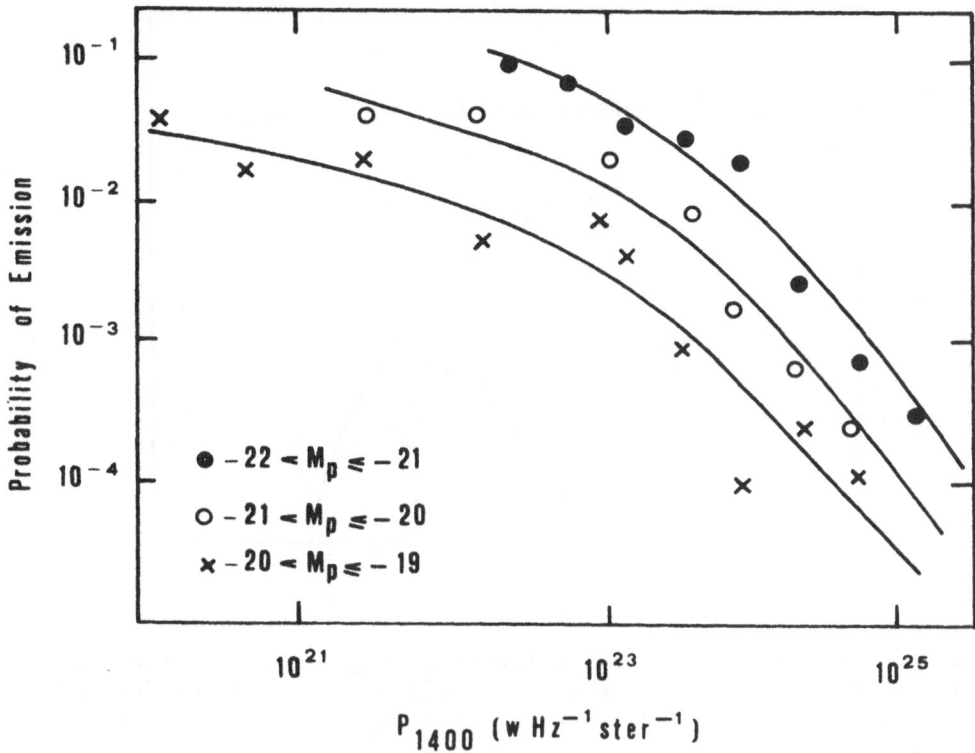

Fig. 4. Bivariate radio luminosity function for elliptical galaxies.

Even if this straightforward interpretation is incorrect, any more detailed model of the time evolution of radio galaxy power must still be consistent with the observed luminosity function.

5.4 The dependence on the optical luminosity

It is well known that radio galaxies are intrinsically very bright optically and have little dispersion in their absolute optical magnitudes. In fact, the dispersion is sufficiently small that apparent magnitudes have been used to estimate distances. To investigate this further, it is necessary to determine the bivariate rather than the monovariate radio luminosity function (Colla et al. 1975). A sketch of the bivariate radio luminosity function is shown in Fig. 4. This has been determined over a wide range in radio power by combining the 3C and B2 radio luminosity functions (Colla et al. 1975), the radio luminosity functions of galaxies in clusters (Jaffe and Perola 1975) and the radio luminosity function of nearby galaxies (Ekers et al. in preparation).

The displacements between the curves for different optical luminosity can be described by a power law $\rho(P_\nu, L_{opt}) \propto (L_{opt})^{-1.5}$. As shown in Fig. 5, it is now clear that the narrow dispersion

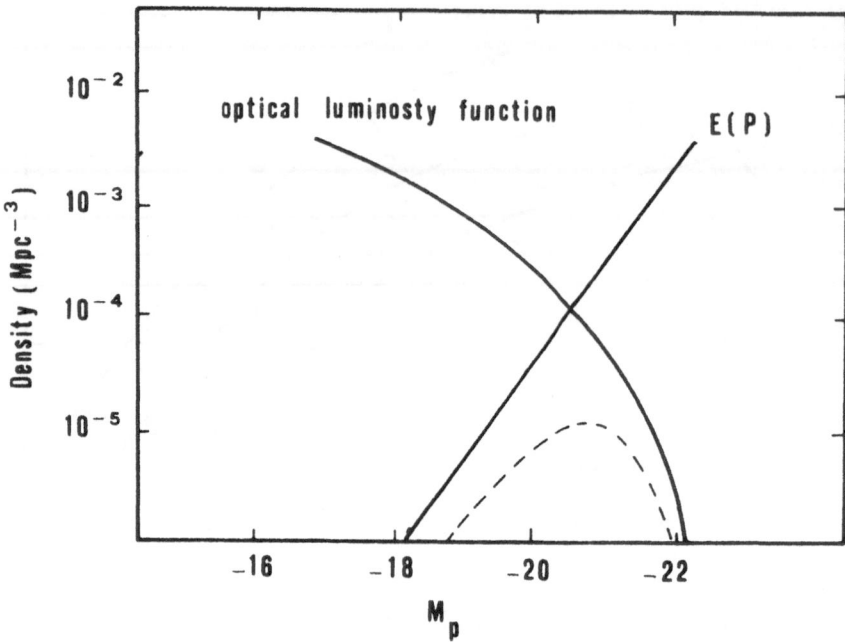

Fig. 5.   The optical luminosity function of galaxies and the probability, E (P), that a galaxy is a radio source.   The product of
these functions gives the distribution of M_p for radio galaxies of
power P,----.

seen in the optical luminosity of radio galaxies results from the
product of the optical luminosity function with the probability
function for radio emission of a given power.

5.5 Cosmology

Radio galaxies form excellent cosmological probes as they can
still be easily detected at distances where the cosmological effects
dominate.   In order to interpret any pure radio test (e.g. log N –
log S) the radio luminosity function is required in order to relate
the observed flux densities to a distance scale.   One important
cosmological test for which it is important to know the bivariate
radio luminosity function is the redshift–magnitude relation using
radio galaxies as standard candles (Sandage 1972).   In Fig. 5 it
was shown how the narrow dispersion in the optical luminosity
arises, but it can also be seen that the average optical luminosity,
M_{pg}, will in fact vary depending on the power of the radio galaxies.
Such variations can be predicted from the bivariate radio luminosity
function and the optical luminosity function, and these should be
included in the analysis.   This is analogous to the Scott (1957)
effect which occurs when using the brightest member of a cluster
as a standard candle.

REFERENCES

Cameron, M.J., 1971, Mon. Not. Roy. Astron. Soc. 152, 429.
Caswell, J.L. and Wills, D., 1967, Mon. Not. Roy. Astron. Soc. 135, 231.
Colla, G., Fanti, C., Fanti, R., Gioia, I. and Lari, C., 1975, Astron. & Astrophys. 38, 209.
Ekers, R.D. and Ekers, J.A., 1973, Astron. & Astrophys. 24, 247.
Heeschen, D.S. and Wade, C.M., 1964, Astron. J. 69, 277.
Hoyle, F. and Burbidge, G.R., 1970, Nature 227, 359.
Jaffe, W.L. and Perola, G.C., 1975, Astron. & Astrophys. (in press).
Longair, M.S., 1966, Mon. Not. Roy. Astron. Soc. 133, 421.
Merkelijn, J.K., 1971, Astron. & Astrophys. 15, 11.
Sandage, A., 1972, Astrophys. J. 178, 25.
Schmidt, M., 1966, Astrophys. J. 146, 7.
Schmidt, M., 1968, Astrophys. J. 151, 393.
Scott, E.L., 1957, Astron. J. 62, 248.

MODELS OF EXTRAGALACTIC RADIO SOURCES

P.A.G. Scheuer

Mullard Radio Astronomy Observatory, University of
Cambridge, Cambridge, England

1. INTRODUCTION

In these lectures I shall first review the basic problems of the
energy supply to radio components and their containment (cf.
Longair, Ryle and Scheuer 1973). Then I want to talk about
models in which energy is transferred from the nucleus to the
radio components by a relativistic beam (Rees 1971, Blandford
and Rees 1974), and about the radio structures to which such
models might lead (Scheuer 1974). I shall show some pictures of
radio sources, which will illustrate how well or badly they fit
the predicted structures, and indicate what modifications we
should try to make in the models. Finally, I shall discuss an
unsolved problem: can we make a relativistic beam stable enough
to get from the nucleus to the extended components without
spreading out over a broad angle?

2. SUPPLYING ENERGY AND MASS TO RADIO COMPONENTS

As Dr. Kellermann said, most radio galaxies – and many
quasars too – are big radio sources around little galaxies, in
fact around even smaller nuclei, for the basic energy generation
seems to happen in the nuclei only. I shall make the
conventional assumption that the radiation (at least that from
the extended components) is synchrotron radiation, and we can
then get a lower limit for the energy in fast particles and
fields from the observed size and radio luminosity of each radio
component; these estimates range up to 10^{51} or 10^{52} Joules.

G. Setti (ed.), The Physics of Non-Thermal Radio Sources, 93–106. All Rights Reserved.
Copyright © 1976 by D. Reidel Publishing Company, Dordrecht-Holland.

(The minimum energy arguments are the same if the radiation is 'synchro-Compton' or 'non-linear inverse Compton' radiation). If there were only fast particles and magnetic fields in the radio components, they would spread out at speed $c/\sqrt{3}$, so one's first instinct is to imagine the fast particles and the fields embedded in clouds of plasma. But then one immediately strikes a difficulty. If a plasma cloud containing fields and fast electrons is ejected from the nucleus (diameter ~ 1 pc, as VLBI measurements show), and expands to form an extended radio component (diameter typically 10 kpc), the energy of each relativistic particle diminishes as radius^{-1} and the magnetic field diminishes as radius^{-2}. The result is that the radio emission should diminish as $(\text{radius})^{-(4\alpha+2)}$. Taking a spectral index $\alpha = 0.75$ and a 10^4 - fold increase in radius, we find that there ought to be compact radio sources 10^{20} times brighter than the extended sources, which is obviously silly. So our first problem is: How do we get energy out of a small energy source into a large radio component, without losing nearly all the energy on the way (mostly by turning it into kinetic energy of expansion of the radio component)? The moral that emerges from a discussion of this problem is that the energy must go out in some form that does not exert pressure, and electrons must be accelerated and magnetic fields manufactured within the extended radio components. I know of four ways of achieving the desired result:

(i) to use the kinetic energy of a moving gas cloud. Interaction with an intergalactic medium may produce shocks or turbulent motion in which particles could be accelerated, and magnetic fields could be amplified by dynamo action.

(ii) to send out beams of relativistic things only, unencumbered with cool gas; the kinetic energy of the relativistic motion is then turned into disordered forms at the end of the beam, where it hits the intergalactic medium.

(iii) to send out machines which are capable of accelerating particles.

(iv) to accept the adiabatic losses inherent in the original plasma cloud picture, but to send out a large number of little clouds in succession, rather than one big one. Then each little cloud only needs to expand by a moderate factor, so that the energetics become more reasonable, and we do not have to account for the collimation of a huge amount of energy all at once, in some impossibly strong nozzle.

The second classic problem is how a radio component can be held together long enough to travel a long distance (sometimes > 50 of its own diameters) from the nucleus. The most obvious ways of stopping or slowing down the expansion of the components use gravitational binding and the inertia of the gas associated with the radio component; it turns out that either of these would require very large masses in the case of really extensive radio components, and very large densities in the case of the

most compact radio components. A third possibility was worked
out by De Young and Axford (1967). In their model, one still
pictures the radio component as a cloud of gas ejected from the
nucleus; as it pushes its way through the intergalactic medium
a bow shock is formed, and the internal pressure of the radio
component is balanced by that of the hot shocked gas within the
bow shock. Since the radio component is continuously
decelerated, the gas within it experiences an apparent
gravitational force in the direction of its motion, and this
limits the extent of the radio component along the axis of the
radio source. Though models of De Young and Axford's type need
less mass in the radio components than the other two, they still
need enough to cause difficulties. Thus, they tend to predict
more Faraday rotation within the radio component than is
consistent with the observations of polarized radio emission.
(See Hargrave and Ryle, 1974 and Longair, Ryle and Scheuer, 1973
for discussions of the case of Cygnus A). However, the greatest
difficulty arises when one considers the supposed ejection of a
large mass of plasma from the nucleus. If the mass is ejected
all at once, most of the total energy of the source must be
released suddenly (for, in these models, the kinetic energy of
the components is much greater than the energy in fast particles
and fields), and extremely strong nozzles would be needed to
direct the flow into two well-defined directions. One would
also require some quite separate energy supply later, to produce
the fast particles and fields; thus this picture fits in best
with energy supply schemes (i) or (iii). The demands on the
collimating mechanism can be reduced somewhat by supposing that
matter is ejected continuously, though with a time scale much
shorter than the age of the source, but one must still invent
some mechanism that gives all the gas nearly the same velocity
V. What one cannot plausibly do is to suppose that gas is
ejected continuously over a time scale comparable with the age
of the source, the gas ejected later catching up with that
ejected earlier. The gas clouds ejected later would impinge on
those ejected earlier with a relative speed comparable with V,
so that the thermal speeds in the radio components, and hence
their speed of expansion, would also become comparable with V.
This argument rules out most versions of scheme (iv) for
supplying the fast particles and fields.

In view of all these difficulties, it is tempting to give
up the attempt to hold together the rather compact "heads" of
radio components, and instead to replace them as fast as they
disperse. That is what happens in Rees' (1971) model. Rees
proposed that pulsars (or things like them) in a nuclear disc
radiate low-frequency electromagnetic waves; the wave frequency
is below the plasma frequency of even the intergalactic medium,
so their radiation pressure must force gas out of the way, and
they excavate a tunnel for themselves; the tunnel walls reflect
the waves at almost glancing incidence, and thus act as a

waveguide. At the end of the tunnel they strike the intergalactic gas, and, being strong waves, they accelerate particles
to high energies. These fast particles radiate at radio
frequencies as they pass through the fields of the low-frequency
waves (by the synchro-Compton mechanism) and then escape.
Blandford and Rees (1974) make much more detailed proposals
about the initial collimation of the beams; Professor Rees will
speak about this "twin-exhaust model" later. They now favour
a beam of relativistic particles rather than a beam of low-
frequency waves, but that makes very little difference to the
dynamics of the outer parts of the source.

   A question that was left unanswered was: "Where do the fast
particles go when they have passed through the end of the beam?"
They cannot use up all their energy in one passage through the
"radio component" of the model, i.e. the end of the beam. The
beam must replace the whole energy in the radio component in a
time d/c, and that energy is at least the minimum energy $U_{min} =$
constant $\times P^{4/7}d^{9/7}$ computed from the radio power P and
diameter d of the radio component, according to the theory of
the synchrotron (or synchro-Compton) process. Therefore

$$\frac{\text{Power radiated}}{\text{Power supplied}} < \frac{\int 4\pi P\, d\nu}{\text{constant} \cdot P^{4/7}d^{2/7}c} \propto \frac{P^{3/7}}{d^{2/7}}$$

Even for the best combination of large P and small d we can find
(such as the "heads" of the radio components of Cygnus A) this
ratio is less than unity, so it is $\ll 1$ in other cases. Thus
great quantities of energetic particles must escape. Presumably
they cannot interpenetrate with the surrounding plasma, so they
must blast it away, forming a cavity full of relativistic
particles (Scheuer 1974). Using some fairly crude approximations,
such as uniform pressure in the cavity (sound speed in cavity
$\simeq c/\sqrt{3} \gg V$), uniform external gas density, and constant power
in the beam, one can make estimates of the size, shape, and
energy density that such a cavity should have. The cavity turns
out to be cigar-shaped -· which is no surprise, since the
pressure at the end of the beam is much greater than that in the
cavity, so the speed V of expansion lengthways is much faster
than the sideways expansion. Much more interesting is the
energy in the cavity. In the purest version of the model, there
are only energetic particles in the cavity, and they need have
no observable effects, but if an appreciable fraction of the
energy appears as magnetic field energy it turns out that the
cavity produces more radio emission than the radio component
at the end of the beam. The exact form of the result depends
a little on how the beam spreads out; the kind of result one
obtains is (Model B of Scheuer 1974)

$$P_{\text{"radio component"}}/P_{\text{cavity}} \simeq 4 \ (V/c)^{1 \frac{1}{8}}(d/D)^{1/4}$$

if the energy is equally divided between magnetic fields and fast particles. Since V < c and d < D, the ratio must be < 4.

I want to mention one more way of helping to hold radio components together. Some clusters of galaxies are X-ray sources, and if the X-rays are bremsstrahlung from hot gas in the cluster one requires densities of the order of 10^3 electrons (and protons) per m^3 and temperatures approaching 10^8 K, and hence a pressure equivalent to a relativistic particle energy density of $\sim 3.10^{-12}$ J m^{-3}. When we compare this with the minimum energy densities for various radio sources:-

Heads of Cyg A components	$\sim 10^{-9}$ J m^{-3}
Tails of Cyg A components	$\sim 10^{-10}$ J m^{-3}
Outer components of 3C66, 3C236, etc.	$\sim 10^{-14}$ J m^{-3}

we see that large diffuse radio components could simply be in pressure equilibrium with their surroundings. The "cavities" described above would then no longer send strong shocks into the surrounding medium as they expand, and if the energy supply from the relativistic beam were interrupted or ceased, they would continue to rise slowly out of the cluster, like the bubbles of Gull and Northover's (1973) model.

3. SOME FEATURES OF THE STRUCTURE OF REAL RADIO SOURCES

Now let us look at some pictures. (Only two of these are reproduced here; most of those shown were from Pooley and Henbest (1974) and Riley and Pooley (1975)). The particular features to which I want to draw attention are:-
(i) there is no noticeable correlation between radio luminosity and the linear size of the source, but there is a good correlation between radio luminosity and the form of the structure (Fanaroff and Riley 1973): sources stronger than 10^{-25} w Hz^{-1} steradian^{-1} (e.g. Cyg A, 3C390.3, 3C430) have the familiar double structure, with the greatest surface brightness at the ends of the radio components furthest from the nucleus (the "heads"), while weaker sources (e.g. Cen A, Virgo A, 3C66, 3C465; also 3C129) have the highest surface brightness near the nucleus, with increasingly diffuse structures further out. Some of the diffuse outer components have very steep radio spectra, and the most natural interpretation is that they are very old sources ($\sim 10^8$ years) in which the particle energy distribution has been affected by synchrotron radiation loss and inverse Compton scattering on the microwave background. That does not explain the surprisingly sharp dividing line in radio luminosity!

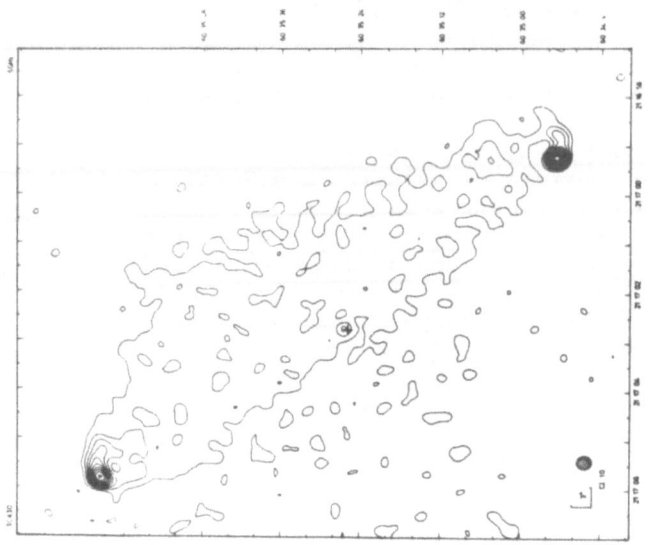

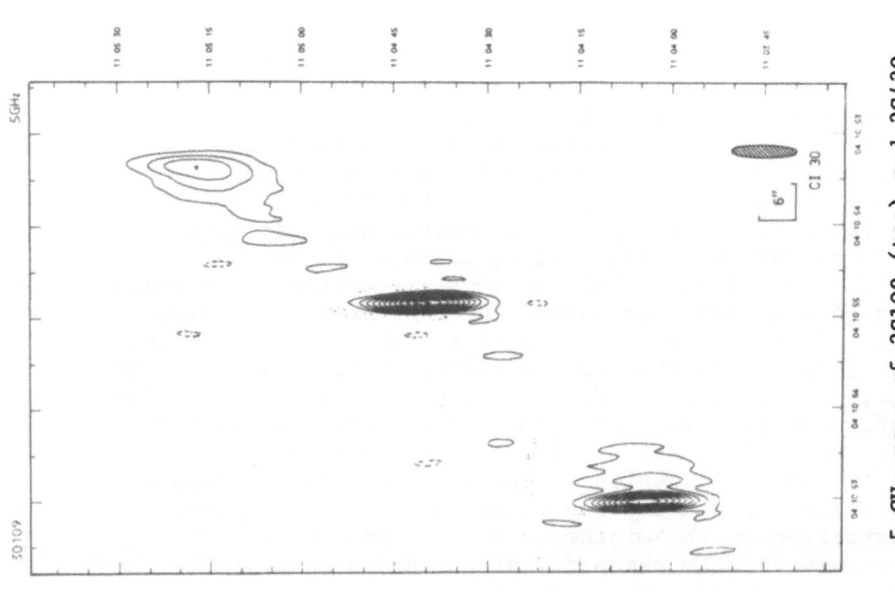

5 GHz maps of 3C109 (top) and 3C430 (bottom) made with the Cambridge 5 km aperture synthesis telescope. Angular resolution 2" x 2"cosec δ.

From Riley and Pooley (1976); reproduced by kind permission of the authors and the Royal Astronomical Society.

(ii) At the high frequency and better angular resolution (2 arc sec) of the new maps, many more sources (about 50%) are showing compact radio components coincident with the nuclei. This observation makes the idea of a continuous supply of energy from the nucleus much more plausible.

(iii) Some sources show bridges of radio emission extending most or all of the way from the "heads" of the radio components to the region of the nucleus. The most beautiful example is 3C430, but there are many others (e.g. 3C 79, 234, 268.2, 284, 452).

Many of these sources have all the features expected in a beam model: the compact nuclear source, the bright "heads" at the furthest ends of the source, and the cigar-shaped "cavity" joining them. On the other hand, many of the radio components have "tails" whose surface brightness decreases gradually with distance away from the "head", and that is incompatible with the simple model of a "cavity" filled uniformly with radio-emitting material. The gap between the "tails" cannot be due to synchrotron losses alone; although there is some tendency for "tails" to have steeper radio spectra than "heads", one observes nothing approaching the striking increase in the length of the "tail" at lower frequencies that one should find if the gap were caused only by energy losses suffered by the older electrons near the centre of the cavity.

While we have the observations before us, I also want to mention the "sling-shot model" of Saslaw, Valtonen and Aarseth (1974). They suppose that massive objects congeal in the nucleus of a galaxy. As long as there are only two of them, they could move in stable orbits around each other, but when there are three, sooner or later two of the objects become tightly bound, leaving enough kinetic energy to separate them from the third and fling the binary and the single object out of the galaxy in opposite directions. Thus they have found a rather natural way in which particle-accelerating machines might be thrown out of a galaxy in exactly opposite directions. It is not at all easy to invent suitable massive objects (Rees and Saslaw 1975). A neutron star cannot supply enough energy, while rotating supermassive objects would be much too luminous to escape detection in the optical band. A black hole even of galactic mass could not accrete enough intergalactic matter to supply the energy required for the radio source, and so far the only suggestion that is allowed by all the constraints that black holes are ejected complete with accretion discs.

In the sling-shot model the radio components would tend to be thrown out in the plane normal to the rotation axis of the nucleus, while in most other models the radio components lie close to the rotation axis. In principle, this is a crucial test, but in practice one cannot see the structure of the nucleus directly, so one has to compare the projected radio axis with the projected rotation axis of the galaxy as a whole. Even

then, the evidence is sparse. The clearest case is Centaurus A;
the galaxy NGC 5128 has an obvious dust lane, which is roughly
perpendicular to the radio axis and also to the rotation axis
indicated by optical spectroscopy. Cygnus A may be a similar
case: the associated galaxy appears double on photographs (the
structure which was at one time interpreted as two colliding
galaxies), but the compact central radio component (Hargrave
and Ryle 1974) lies between the two optical images. These
observations suggest that the image of the galaxy is split by
a dust lane, which obscures the nucleus. With that
interpretation, the projected radio and rotation axes are
within 5° of each other. Mackay (1971) and Bridle and Brandie
(1973) compared the axes of radio sources with the axes of the
associated elliptical galaxies; there may be some tendency for
the radio source to lie along the major axis, but it is not
very significant. Though some hundreds of radio sources have
now been identified with E galaxies, most of these galaxies are
rather faint, and most of them are very round E galaxies, and
it is surprisingly difficult to improve the statistics. In
addition, some galaxies (e.g. 3C33) have inner parts whose long
axis lies in a different P.A. from that of the outer, more
diffuse parts. We must also bear in mind that the longer
optical axis of Centaurus A lies along the rotation axis.

4. COMMON GROUND

At this stage I want to summarise those features that most
acceptable radio source models have in common.
(i) The acceleration of electrons to high random velocities
take place within the radio source components. General
arguments about adiabatic losses led us to this conclusion, and
in a few cases it has fairly direct support from observation.
If the optical continuum from the jet of M87 (and 3C273) is
synchrotron radiation, the lifetimes of the relevant electrons
are much shorter than the light travel time from the nucleus.
(ii) The head-tail structure of the components of powerful
sources shows that there is some interaction with an intergalactic medium. Without such interaction, the radio emitting
material would spread out more or less isotropically in the
rest frame of the "head" of the radio component; it would not be
left behind like a trail of smoke.
(iii) The matter density within the radio source components is
large enough to affect their dynamics. The systematics
depolarization of radio sources at low frequencies, attributed
to Faraday rotation, indicates matter densities likely to have
important dynamical effects. We also found that, if one
neglects the matter density, one predicts radio structures that
resemble some sources such as 3C 430, but which do not agree

with the more typical head-tail structures of radio components.
       The story that emerges most naturally is that fast
particles and magnetic fields first appear in the "head" of a
radio component: their energy density u gives rise to a
pressure 1/3 u which has to be balanced by the ram pressure
ρ_{ext} V^2 of the intergalactic medium. It seems natural to
continue that the fast particles then flow out of the high-
pressure "heads" into the lower-pressure "tails", but that
simple statement would be wrong. It is well-known in hydrodynamics that any steady flow along diverging streamlines must
travel from lower pressure to higher pressure unless the flow
is supersonic relative to the pattern of streamlines. So either
we have a flow which is supersonic throughout - even within the
"heads" - or we must have a constriction (a de Laval nozzle
again) near the head in which the flow can change from subsonic
to supersonic. A third possibility is that we do not have a
steady streamline flow at all, but a turbulent flow which
entrains some of the surrounding intergalactic matter; it may
be that in this case a subsonic flow is allowed. This third
possibility appeals to me most, because we saw earlier that the
transport of matter from the nucleus to the radio components
leads to difficulties.
       Evidently one ought to be able to understand a good deal
about the dynamics of radio components without choosing any
particular theory of the origin of the fast electrons. The way
in which the energy appears in the "heads" affects the cross-
section of the "head", and hence the rate at which the "head"
advances through the intergalactic medium, but that is the only
way in which it enters into dynamical calculations. Nobody has
done the hydrodynamics yet, and it will not be easy. The
results would probably not tell us much about the origin of the
energy, but they might at least show whether we are thinking
along the right lines.

5. THE STABILITY OF BEAMS

If one squirts a jet of coloured water into clear water, it
spreads out at an angle of about 15° on each side of its mean
direction (Landau and Lifshitz 1959, p. 133). Such a jet is
obviously no use for supplying energy from the nucleus to radio
source components; it would make the "heads" of radio components
much too large, and besides, the mixing with surrounding matter
would bring back the adiabatic losses that one has to try so
hard to avoid.

radio source
model

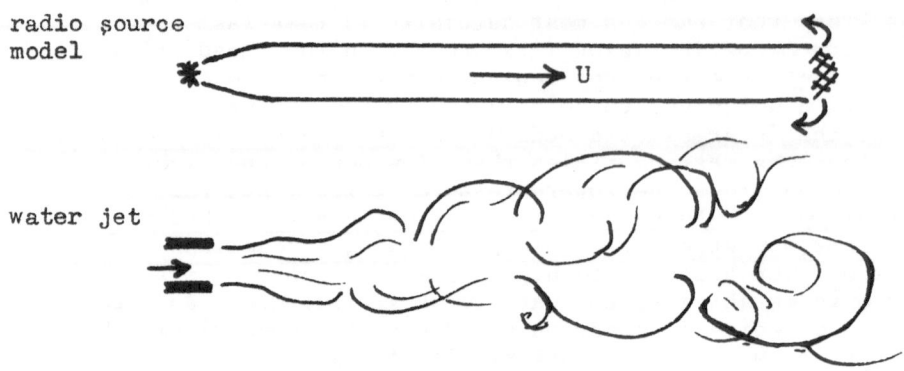

water jet

Why must such a jet spread out? An unperturbed jet would be
surrounded by a tangential velocity discontinuity (a vortex
sheet), and growing modes of surface waves exist around such a
discontinuity (the Kelvin-Helmholtz instability). The growth
is very fast in the linear regime; for real wave-vector $\underline{k}$ the
real and imaginary pars of ω are about the same (in fact they
are equal if we have the same incompressible fluid on both
sides of the vortex sheet). An oversimplified picture of
events is this: waves of some short wavelength λ grow to an
amplitude $\lambda/2$ in a few e-folding times, a time $2\pi/\omega$ say, so the
fluids are then mixed through a shear layer $\lambda/2$ wide, in a time
in which layer has been convected through a distance $(U/2)(2\pi/\omega)$
λ. Waves shorter than the thickness of the shear layer no
longer grow, but by the time the layer has travelled 2λ, waves
of twice the original wavelength have had time to grow, so
bigger eddies appear - and so on.

What B.D. Turland and I have tried to find out is whether
the instability can be made slower or suppressed if one has a
relativistic jet. For our first attempt we have only looked
at the linear regime, and we took the simplest possible
geometry, which is a plane vortex sheet. For two identical
incompressible fluids, surface corrugations with wave number k
(real and parallel to U) have $\omega = \frac{1}{2}kU(1 \pm i)$, so the growth (or
decay) is $e^{2\pi}$ per period. Now suppose that the fluids are
compressible, that is, the sound speeds $a_1 = (\Gamma_1 P/\rho_1)^{1/2}$ and
$a_2 = (\Gamma_2 P/\rho_2)^{1/2}$ in the two fluids are finite. We expect little
change so long as both sound speeds are $\ll U$, the relative
velocity, but we can expect drastic changes in behaviour when
$U > a$, i.e. when the flow becomes supersonic. The dispersion
relation for surface corrugations perpendicular to $U(\underline{k} \parallel U)$
is

$$\frac{(k^2-\omega^2/a_1^2)^{\frac{1}{2}}}{\Gamma_1(\omega^2/a_1^2)} + \frac{(k^2-(\omega-kU)^2/a_2^2)^{\frac{1}{2}}}{\Gamma_2(\omega^2/a_2^2)} = 0$$

(with the rule that one must take the square roots with non-negative real parts, in order to exclude waves that become large at great distances from the vortex sheet). It is clear from the dispersion relation that there are no real solutions for ω and k if the relative motion is subsonic; with a bit of ingenious algebra one can show that, when $\Gamma_1 = \Gamma_2$, unstable modes exist so long as

$$U < (\ a_1^{2/3} + a_2^{2/3}\ )^{3/2}.$$

When U is above this limit, the genuine solutions of the dispersion relation merely represent sound waves which are incident at such an angle that they are refracted without reflection. However, oblique surface waves can still grow provided only that

$$U \cos \theta < (\ a_1^{2/3} + a_2^{2/3}\ )^{3/2}$$

where θ is the angle between $\underline{k}$ and $\underline{U}$ (Miles 1958, Fejer and Miles 1963).

Can one suppress the instabilities if U and perhaps also the sound speeds a_1 and a_2 are relativistic? Deriving the dispersion relation is surprisingly easy; corresponding to the non-relativistic equation above, one obtains

$$\frac{(k_1^2-\omega_1^2/a_1^2)^{\frac{1}{2}}}{\Gamma_1'(\omega_1^2/a_1^2)} + \frac{(k_2^2-\omega_2^2/a_2^2)^{\frac{1}{2}}}{\Gamma_2'(\omega_2^2/a_2^2)} = 0$$

where $\Gamma' = \Gamma + a^2/c^2$; $(\underline{k}_1, \omega_1)$, $(\underline{k}_2, \omega_2)$ are the $\underline{k}$ and ω of the wave in the rest frames of fluid 1 and 2 respectively, and they are related by the usual Lorentz transformation. While it is easy to derive, this dispersion relation is messy to solve. But it is still quite simple to show that sufficiently oblique surface corrugations are always unstable; in fact, if a_1, a_2 are both $\ll$ c, condition for instability is just

$$U \cos \theta < \{\ a^{2/3} + (a/\gamma)^{2/3}\ \}^{3/2}, \qquad (\gamma^{-2} = 1 - U^2/c^2)$$

and if we make θ close enough to $\pi/2$ this can always be satisfied. When one or both sound speeds are relativistic, we have to use numerical solutions, but the results are qualitatively similar. For highly supersonic U the biggest growth rate occurs for values of U cos θ just below the critical value; for large γ the maximum growth varies like $\gamma^{-2/3}$.

We can easily understand why very oblique perturbations

remain unstable for all U. We have assumed that the fluids are
inviscid, so we can imagine them to be separated by an
infinitely slippery polythene film. The boundary conditions
remain the same, but each fluid feels only the displacements of
the other fluid perpendicular to the (unperturbed) interface,
so the fluid behaves as if the other fluid moved at speed U cosθ
along **k**. The sliding motion along the surface corrugations
has no effect.

So far we have treated the relativistic beam as a strictly
inviscid fluid, i.e. a gas whose particles have exceedingly
small mean free paths. A small viscosity immediately makes the
problem much more complicated, but one can deal with the other
extreme, when the mean free path in one fluid is very large, as
it might be if one "fluid" is a beam of low frequency photons.
The model we adopted is this:

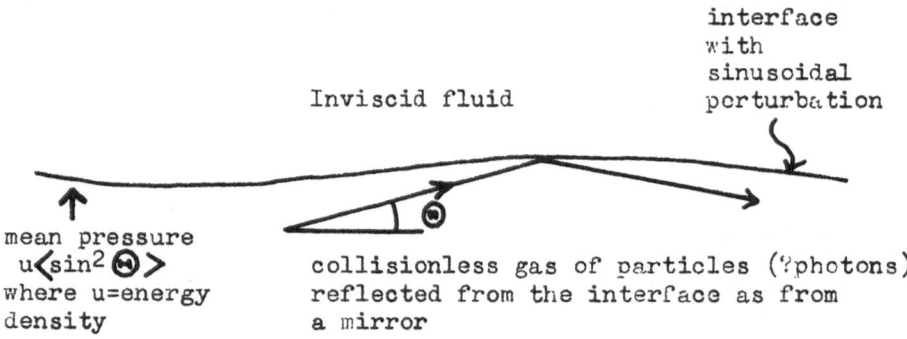

interface
with
sinusoidal
perturbation

Inviscid fluid

mean pressure
u$\langle$sin^2 Θ $\rangle$
where u=energy
density

collisionless gas of particles (?photons)
reflected from the interface as from
a mirror

The dispersion relation for this situation looks quite
different from that for two inviscid fluids, but the numerical
results are fairly similar if one replaces $1/\gamma$ with a parameter
that is an average value of Θ. In particular, there is always
an instability for sufficiently oblique perturbations.

Taking the simplest view, we might suppose that a beam of
diameter d will start to break up when perturbations of wavelength d have had enough time to grow. Typically, a beam has to
travel $\sim$ 30d from the nucleus, if we identify d with the diameter
of the "head" of a radio component. A relativistic beam travels
at speed U $\simeq$ c, so the time available for growth is $\sim$ 60d/c. If
this is to be $\leq$ 2 e-folding times for k = 2π/d, we require the
imaginary part of the phase velocity (ω/k) of the waves to be
$\leq$ 0.005c. To bring the maximum growth rate below this value
requires $\gamma \gtrsim$ 50 if the sound speed is 0.1c in both the beam and
its surroundings, and γ = several hundred if the sound speeds
are c/$\sqrt{3}$. Thus it is possible to bring the formal growth rates
of the instability down to tolerable levels if one can accept
rather large values of γ for the beam.

Having said this, I want to add a few cautionary remarks.

(i). All the above is linearised analysis. In most problems
of this sort one is more interested in the way the system
behaves when it has got into a thoroughly non-linear regime, and
often the linear analysis has little relevance to the final
result. However, in this case the linear analysis may be
useful, because if a beam mixes with the surrounding gas it
loses its purpose in radio source models, and we are not very
interested in its later development.
(ii). Small scale (large k) perturbations grow faster than
perturbations with $\lambda \sim d$. The kinetic energy of the relative
motion of the gases in the mixing layer will decay through
turbulence into heat, and if the relative motion were highly
supersonic this heat input would be much greater than the
initial thermal energy. The temperature and pressure of the
mixing layer would then rise sharply, and it would very
probably drive a shock which would heat the surrounding medium.
Thus a highly supersonic value of U is probably an unrealistic
situations, and we ought not to appeal to highly supersonic
flow to reduce the rate of growth of instabilities.
(iii). Our analysis gives the growth rate (imaginary part of ω)
for real k. When a wave grows as it progresses along a beam, we
ought to think about complex k and real ω. One cannot properly
derive the latter situation from the former using a group
velocity when the growth rate is large (see e.g. Mattingly and
Chang 1974).
     So we do not know yet whether one can invent a sufficiently
stable beam; we only have the first step towards the physics
that we really need.

REFERENCES

Blandford, R.D., and Rees, M.J., 1974, M.N.R.A.S. 169, 395.
Bridle, A.H., and Brandie, G.W., 1973, Astrophys. Lett. 15, 21.
De Young, D.S., and Axford, W.I., 1967, Nature Lond. 216, 129.
Fanaroff, B.L., and Riley, J.M., 1974, M.N.R.A.S. 167, 31P.
Fejer, J.A., and Miles, J.W., 1963, J. Fluid Mech. 15, 335.
Gull, S.F., and Northover, K.J.E., 1973, Nature Lond. 244, 80.
Hargrave, P.J., and Ryle, M., 1974, M.N.R.A.S. 166, 305.
Landau, L.D., and Lifshitz, E.M., 1959, Fluid Mechanics (Pergamon
Press).
Longair, M.S., Ryle, M., and Scheuer, P.A.G., 1973, M.N.R.A.S.
164, 243.
Mackay, C.D., 1971, M.N.R.A.S. 151, 421.
Mattingly, G.E., and Chang, C.C., 1974, J. Fluid Mech. 65, 541.
Miles, J.W., 1958, J. Fluid Mech. 4, 538.
Pooley, G.G., and Henbest, S.N., 1974, M.N.R.A.S. 169, 477.
Rees, M.J., 1971, Nature Lond. 229, 312.
Rees, M.J., and Saslaw, W.C., 1975, M.N.R.A.S. 171, 53.
Riley, J.M., and Pooley, G.G., 1976, Memoirs R.A.S. in press.

Saslaw, W.C., Valtonen, M.J., and Aarseth, S.J., 1974, Ap. J.
190, 253.
Scheuer, P.A.G., 1974, M.N.R.A.S. 166, 513.

BEAM MODELS FOR DOUBLE SOURCES AND THE NATURE OF THE PRIMARY
ENERGY SOURCE

M.J. Rees

Institute of Astronomy, Cambridge, England

1. INTRODUCTION

In these lectures I shall first summarise some aspects of the
so-called "beam" or "twin-exhaust" model for double radio sources;
and then comment on the emission mechanism for non-thermal radiation
from the nuclei of galaxies, and the nature of the central energy
source.

2. DOUBLE SOURCES

2.1 The basic model

General critiques of radio source models are given by Longair, Ryle
and Scheuer (1973), and Blandford and Rees (1976). I restrict
attention here to some implications of the particular class of theory
discussed in the following references: Rees (1971), Blandford and
Rees (1974), Scheuer (1974) and Rees (1975). The reader is referred
to these papers for basic descriptions of this model, according to
which the energy content of extended sources is continuously supplied
and replenished by beams of fast-moving plasma emanating from the
nucleus of the associated galaxy. The model's essential ingredients
are therefore:
 A power supply in the galactic nucleus which produces an outflow of magnetised plasma. This plasma may be relativistic (in the
sense that $p/\rho \simeq c^2/3$), but all that is actually necessary is that
p/ρ should exceed V^2, where V is the escape velocity from the central
gravitational potential well.
 A mechanism for bifurcating the outflow and collimating it into
two oppositely-directed beams. One such mechanism, involving the

formation of "nozzles", was proposed by Blandford and Rees (1974).

    <u>An acceleration mechanism in the radio components.</u> In these
models, the high-surface-brightness regions of the radio blobs are
identified with the places where the beams are impinging on the
extragalactic medium. By the time the beams have penetrated $\gtrsim 10^5$
pc from the galaxy, most of the relativistic plasma's initial
internal energy will have been transformed (in accordance with
Bernoulli's equation) into bulk kinetic energy. A mechanism is
therefore required for reconverting the energy into random particle
motions, and establishing a power law spectrum for the relativistic
electrons. The actual mechanism is unknown; but the kind of acceleration which almost certainly has to operate with $\gtrsim 1\%$ efficiency
in, for instance, the supernova remnant Cas A (where the velocities
are $\sim 0.01c$) could be even more efficient behind shock fronts where
the velocities are much higher.

    If the flow pattern were sufficiently stable, successive stages
in the evolution of a typical double source might be as shown in
Fig. 1: the "hot spots" move outward, leaving a sheath or cocoon
of lower surface brightness along their track; but eventually the
central source switches off, and the residual relativistic plasma
then expands and merges into the inter-galactic medium. The precise time dependence of source size and radio luminosity depends on
the external density, on how well the beams are collimated, etc.,
etc. If this type of model applies to the "giant" double source
3C 236, the beams must have lasted for $> 10^8$ years.

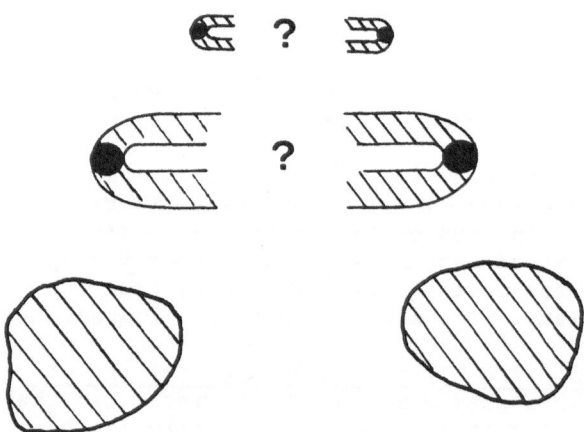

Fig. 1. Qualitative evolution of a double source, showing the
"hot spots" and the cocoon. When the central activity (and therefore the beam) switches off, the "hot spots" disappear and the
components expand and fade.

2.2 Complications

Factors likely to complicate the interpretation of actual sources
include the following:
      i) Kelvin-Helmholtz instabilities at the sides of the beam may
cause entrainment of surrounding material. The beam would then
widen faster than if its flow remained strictly isentropic.
     ii) The central power supply may vary.
    iii) Instabilities at the "nozzles" may make the intensities of
the two beams unequal. One could even envisage "flip-flop" behaviour, where the relativistic outflow squirts alternately in two
opposite directions. This may be relevant to "jets" such as those
in 3C 273 and M87. (In the latter case, it could account naturally
for the greater symmetry of the radio structure (Turland 1975) -
the radio-emitting electrons, having longer synchrotron lifetimes,
may survive for several reversals of the beam direction even if the
optical electrons do not).
     iv) Inhomogeneities in the extragalactic medium may cause complex
structure in the hot spots, as is seen in 3C 390.3 - for instance,
part of the beam may lag behind the rest, or be deflected, if it
encounters a dense external cloud.
      v) Transverse motions of the external medium relative to the
galaxy could destroy the symmetry or linear structure, especially
when the beam is so weak, or so poorly collimated, that its speed
of advance is $\lesssim$ the transverse velocity. This is perhaps relevant
to the interpretation of "radio trails".
      If the beams varied on timescales 10^4 - 10^7 years for the
reasons cited under ii) and iii), the cocoon, which delineates
the path traced out by the hot spots over the history of the source,
would be non-uniform: it would be particularly conspicuous in places
where the hot spots were located at times of high beam intensity.
The cocoon could thus have the appearance of pairs of blobs closer
than the outer hot spots to the central galaxy. (This is perhaps
relevant to the interpretation of so-called "double doubles").

2.3 Setting up the flow pattern

The possibly severe effects of instabilities, as described in
Scheuer's lectures, pose one of the main uncertainties in the "twin
exhaust" model. However, even if the basic flow pattern is stable
enough, we still ought to consider how it can be set up. In practice, the gas cloud around the nucleus may not form until the violent activity starts (or the gas may already be there, but in a
cool thin disk). But the simpler problem of what happens when a
source of relativistic fluid "switches on" at the centre of a pre-
existing oblate spheroidal gas cloud is still unsolved, though it
actually resembles a scaled-down version of the situation discussed
by Gull and Northover (1973). I would conjecture (and this is no
more than a conjecture at the moment!) that the eventual steady
state is established on a fashion that depends on whether the power

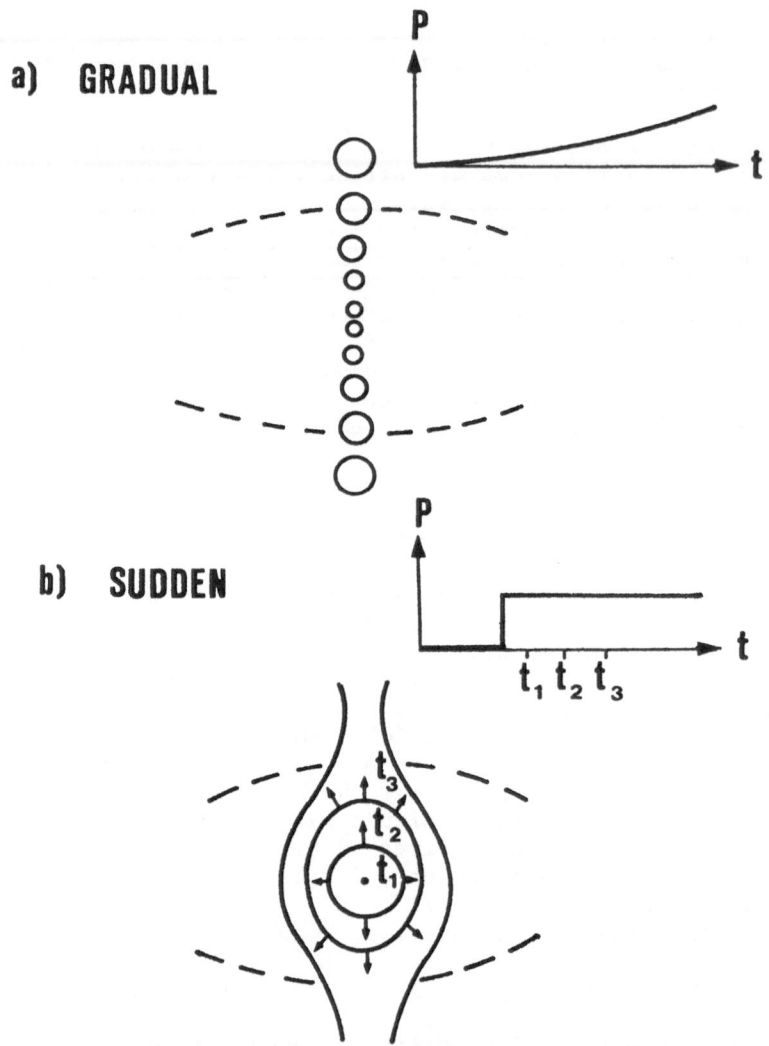

Fig. 2. Two possible scenarios for the formation of beams when a central source of plasma switches on at the centre of an oblate spheroidal gas cloud in a gravitational potential well.

P(t) of the central source switches on gradually or suddenly:
     a) <u>Gradual</u>. While P(t) remains very low, the relativistic plasma escapes in two trails of bubbles, rising under buoyancy along the axis where the pressure gradient is steepest (Fig. 2a). These bubbles rise at a speed of order

$$\text{(sound speed in external gas)} \times \left(\frac{\text{bubble size}}{\substack{\text{scale height of gas cloud} \\ \text{along minor axis}}}\right)^{\frac{1}{2}}.$$

Because this speed is low, there is a limit to the rate at which
the relativistic plasma can be transported away in this manner.
When P(t) exceeds some critical value, successive bubbles merge to
form a continuous tube. The outflow velocity is then $\sim c/\sqrt{3}$ - i.e.
highly supersonic with respect to the external sound speed.
    b) Sudden. If P(t) immediately attains its peak value, then
the bubble of relativistic plasma will typically inflate at a rate
exceeding the external sound speed. Its boundary is then stable
against the Rayleigh-Taylor instability (c.f. Gull and Northover
1973). The bubble then grows to a full scale height without bifurcating, and may then transform directly into a beam as sketched
in Fig. 2b, (c.f. the discussion of explosions in exponential
atmospheres by Kompaneets (1960)).

2.4 Relevant (?) experiments

Studies of the following experimental situations, where a "light"
fluid is injected into a "heavy" fluid in a gravitational field,
may provide some insight into the character and stability of the
flow pattern envisaged in the "twin-exhaust" model:
    i) Release a steady flow of air under water, at a depth equiva-
lent to several pressure scale heights. (This would require a lake
$\gtrsim 30$m in depth. Alternatively, the experiment could be performed
on a smaller scale under reduced pressure in the laboratory).
Conjecturally, the outcome might resemble that shown in Fig. 3: for
low flow rates, the air would rise as a series of separate bubbles;
but for high flow rates a channel would be established, the cross

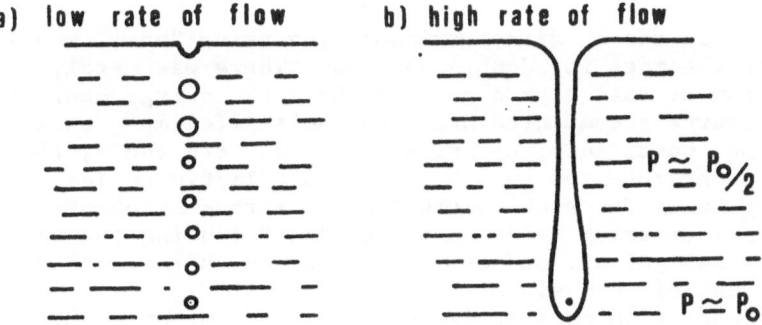

Fig. 3. Conjectural outcome of experiments where air is steadily
released under water at a depth where the total pressure is several
times atmospheric. (Compare with Fig. 2).

section being a minimum at the level where the external pressure is $\sim \frac{1}{2}$ its value at the source.

ii) Experiments on "fluidised beds" - where air is pumped from below into a layer of sand or powder - might indicate how stable jets can be (Spiegel, 1975).

2.5 The applicability of a "fluid dynamical" treatment

Although the collision mean free paths of relativistic particles in radio sources are very long, a fluid dynamical treatment can be justified because we know that magnetic fields are present. The gyroradius of a relativistic particle of energy ε moving in a magnetic field B gauss is $\sim 10^{-12}$ ($\varepsilon/m_p c^2$) B^{-1} pc. This is typically very small compared to all relevant scales for magnetic fields $\gtrsim 10^{-6}$ gauss. For this reason - and also because collective plasma effects may also reduce the mean free path - we adopt a fluid treatment (c.f. the solar wind, which can in many contexts be regarded as a fluid, even though collisional mean free paths are again very long).

If all the particles are ultrarelativistic, then $p = 1/3 \, \rho c^2$ and the sound speed is $c/\sqrt{3}$. Bernoulli's equation is then simply $\gamma_{bulk} = (1 - V^2/c^2)^{-1/2} = (p/p_0)^{-1/4}$; and the mean random energy per relativistic particle, measured in the moving frame, varies as γ_{bulk}^{-1}. If the magnetic field contributes significantly to the total energy density and has a preferred orientation, then the pressure and magnetosonic velocities are, of course, anisotropic. The Debye length is also very small compared with the scale of the flow. This means that the relativistic plasma must be neutral, and the relativistic generalisation of ordinary MHD is applicable. There will be no electric field in a frame sharing the mean plasma velocity V; but in a non-moving frame, of course, the electric and magnetic field energies would be comparable if V ≃ c.

2.6 Is there any evidence favouring the "twin-exhaust" model?

Five years ago there was no direct evidence favouring "beam" models over alternative theories for double sources. There was merely the indirect argument that such a model - where the energy content of an extended source accumulated over its whole lifetime - obviates the need to invoke power supplies exceeding $\sim 10^{46}$ erg sec^{-1}, the characteristic power inferred by other data on galactic nuclei.

But the arguments for such models are now rather stronger:

i) There is increasing evidence of "bridges" linking the radio blobs to the central galaxy, and of continuing non-thermal activity in the galactic nuclei themselves.

ii) Higher resolution maps of Cyg A, 3C 236 and other double sources reveal "hot spots" within the components where the electron lifetime may be shorter than the light travel time from the central galaxy. This implies continuous injection or reacceleration of electrons in the components.

iii) On the other hand, the continued absence of evidence for
milli-arc-second structure seems to argue against the existence
of compact gravitationally-bound particle accelerators there.

iv) Improved upper limits on the gas density in the components
raise difficulties for the DeYoung-Axford (1967) or inertial confinement models.

v) Recent evidence that the small central sources in Cygnus A,
3C 111 and 3C 236 are elongated and aligned with the overall source
axis shows that there is a continuing energy source with a preferred
direction (rotation axis?) which is "remembered" for $> 10^8$ years
(Kellermann et al. 1975, Fomalont and Miley 1975).

Some outstanding observational issues whose elucidation seems
prerequisite for further progress in interpreting extended sources
include:

i) What is the temperature, density and distribution of gas
around the sources? (X-ray maps from HEAO B* with $\sim$ 2" resolution
should help to answer this question.)

ii) Does the non-thermal radio spectrum from the components
extend upward into the optical band? If so, this would place more
stringent demands on the "in situ" particle acceleration mechanisms.

iii) What is the smallest scale structure in the "hot spots"?
Very small scale, i.e. VLBI-type, structure would imply that the
components contained compact objects, or that the radio emission
involved some coherent process more efficient than synchrotron
radiation. Structure on scales $\leq 0.1"$ would set lower limits to
$\rho_{ext} V^2$ which would be embarassing for any ram pressure confinement
model, especially for sources outside clusters.

iv) Do sources inside and outside clusters differ systematically
in a) overall extent, or b) compactness of components?

v) How is source axis aligned with respect to rotation axis
of the galaxy?

vi) How does the polarization and spectral index vary over the
source? This is relevant to the flow or diffusion rate of relativistic particles in the cocoon region, and to the magnetic field
structure. (In "beam" models, the magnetic field in the components
could have been transported out along with the relativistic plasma.
The field direction will then be determined by the flow pattern.
Shearing motions can build the field up to equipartition strength,
but its back-reaction on the flow would prevent it strengthening
further.)

3. EMISSION FROM GALACTIC NUCLEI

3.1 Preliminaries

If the components of extended sources were indeed just clouds of
relativistic plasma, as the "beam" model would imply, this would
really diminish the motivation for delineating their higher-reso-
lution structure in elaborate detail but would enhance the importance

of studying the compact central object that initiates and ener-
gises them. To use a perhaps unsavoury analogy: once one realises
that the mushroom cloud from a nuclear explosion owes its shape
to interaction with a stratified surrounding atmosphere, the cloud's
detailed morphology becomes less physically interesting than the
explosive energy release itself.

The radio data on galactic nuclei are reviewed in the accompanying lectures by Kellermann. The apparent "superluminous"
transverse velocities are often interpreted in terms of the straightforward kinematics shown in Fig. 4. If the radiating region moves
with a Lorentz factor $\gamma \gg 1$, then (whether this is a real physical

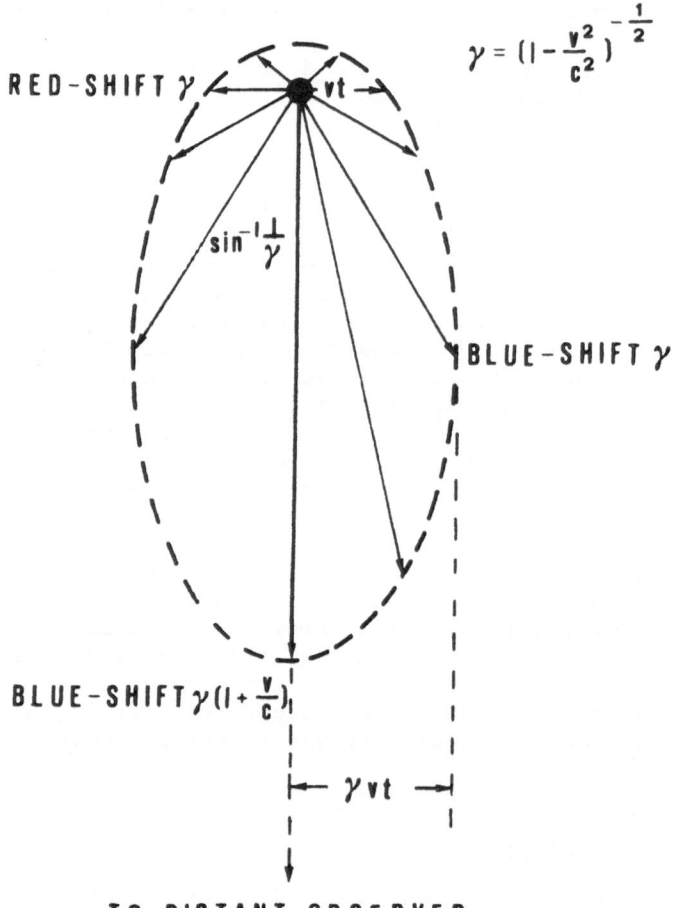

Fig. 4. The kinematics of uniform relativistic expansion with
velocity V. The ellipse represents the apparent shape of the
emitting surface at a time t after the expansion is observed to
start. The apparent transverse velocity can be as large as γV.

velocity or just a phase velocity) apparent transverse motions up
to $\sim \gamma c$ are possible for velocities making angles $\sim \sin^{-1} \gamma^{-1}$ with
the line of sight. It is therefore not easy to interpret apparently
symmetric expanding doubles as two blobs moving in opposite directions. Note, however, that this difficulty is eased if the
motion involves a phase velocity instead of actual relativistic
bulk motion, because only the latter case necessarily entails
large, and very different, Doppler effects for the two components.

3.2 Dynamics of relativistic expansion

It is still uncertain whether the radio emission from compact
sources is really synchrotron radiation at all. While it is
therefore premature to devise detailed synchrotron models for particular sources, it may nevertheless be worth while to describe
how the merely kinematic models conventionally discussed could be
extended to incorporate realistic gas dynamics. The following simple cases of relativistic fluid flow are discussed in the recent
or forthcoming literature:

   a) Free one-dimensional expansion of semi-infinite medium into
a vacuum (Mathews 1971). There is here a similarity solution where
a rarefaction wave propagates inwards from x = o at $c/\sqrt{3}$. The front
edge moves outward at c. An element of fluid initially at −x starts
to move at a time t = $\sqrt{3}$ x/c and is thereafter exposed to a pressure
gradient which gradually accelerates it up towards speed c.

   b) Free expansion of an initially uniform spherical cloud of
relativistic gas (Ozernoi and Ulanovsky 1974, Salvati and Vitello
1975. If an initially uniform cloud of ultra-relativistic plasma
starts expanding from radius r_o, then the Lorentz factor of the
bulk expansion when the sphere has attained a radius r >> r_o, will
be $\sim r/r_o$, the material then being mostly in a thin shell whose
thickness remains $\sim r_o$. The internal energy per particle measured
in a comoving frame, would of course decrease by the same factor
$\sim r/r_o$.

   c) Steady relativistic wind. Bernoulli's equation for a spherically symmetric relativistic outflow requires that the Lorentz
factor for the bulk flow be proportional to r. Thus, each comoving
shell in the wind behaves rather like the expanding sphere in b).
Note that, if such a wind were to make a transition from subsonic
to supersonic flow on passage through a "sonic radius" where the
gravitational field plays the role of the nozzle (by analogy with
the solar wind) then this would need to occur at a place where the
escape velocity were $c/\sqrt{3}$ - i.e. near the Schwarzschild radius of
a relativistically deep potential well.

   d) Relativistic blast wave (Blandford and McKee 1975). If a
"point explosion" causes relativistic expansion of a blast wave
into a uniform surrounding medium, then $\gamma_{blast} \propto r^{-3/2}$. The swept-
up particles acquire random energy, measured in a frame moving
with the shock, corresponding to $\gamma \simeq \gamma_{blast}$. These particles are
concentrated in a shell of thickness $\sim ct\gamma^{-2}(\propto r^4)$. When r gets

so large that the shock stops being relativistic, there is a transition towards the ordinary Sedov solution.

Any of the above could be elaborated into a possible model for variable radio sources. Note, however, that one of the few things VLBI studies tell us unambiguously is that actual compact sources are not spherically symmetrical! Note also that the bulk expansion rate in models a) – c) would in fact "saturate" at a finite γ equal to the mean Lorentz factor of the initial thermal motions. These models are applicable only if the external medium is rarified enough to be ignored. For a relativistic outburst of initial total energy $\mathcal{E}$, expanding with Lorentz factor γ, this requires

$$\frac{4}{3} \pi \rho_{ext} R^3 c^2 \lesssim \mathcal{E}/\gamma^2,$$

i.e. the particle density in the external medium must satisfy

$$n \lesssim 10^{-1} \left(\frac{\mathcal{E}}{10^{54} \text{ ergs}} \right) \gamma^{-2} \left(\frac{R}{10^{19} \text{ cm}} \right)^{-3} \text{cm}^{-3}.$$

Otherwise the relativistic ejecta would be braked at too small a radius to provide an adequate model for the radio structure.

3.3 Radiation mechanisms, etc.

If the radio emission from compact souces involved a coherent mechanism, then high surface brightness temperatures (i.e. T >> >> 10^{12}°K) could be attained, without the well-known constraints imposed by synchrotron self-absorption and inverse Compton losses. There would then be no need to invoke superluminous apparent motions to explain the variability; but "Christmas tree" effects could still of course occur. We might then expect rapid intrinsic low frequency variability, or else (if observations are made with a narrow enough band-width) variations due to interstellar scintillations (Lovelace and Backer 1972).

It is conceivable that a scaled-up version of the pulsar emission mechanism operates in galactic nuclei. Another suggestion (Colgate, Colvin and Petschek 1975 and earlier work cited therein) is that supernova-induced shock waves may excite oscillations at the plasma frequency $\nu_p \simeq 9 \times 10^3 n_e^{1/2}$ Hz. These then give rise to photons with $\nu = 2\nu_p$; and these photons then diffuse upwards in frequency owing to repeated coherent scattering off plasma oscillation. This process becomes inefficient by the time frequencies $\sim 10^{11}$ Hz have been reached; but ordinary Compton scattering off hot (but non-relativistic) electrons may then take over and reproduce the optical and ultraviolet part of quasar spectra. Katz (1975) has computed the resultant spectrum taking account of photon diffusion in real space as well as frequency space, and finds that power law spectra of appropriate slopes can plausibly be obtained. A possibly serious complication, however, is that the brightness temperature may be high enough for induced Compton scattering to be important. This tends to cause a <u>downward</u> drift in the radiation frequency.

3.4 Some general notes on galactic nuclei

The range of tenable options for the primary energy source in active galactic nuclei still seems as wide as it ever was. (See Saslaw (1974) for a recent review.) The main contending models are these:

a) Clusters of stellar mass bodies (i.e. multiple supernovae, multiple pulsars, etc.).

b) Supermassive stars (Hoyle-Fowler (1963) objects, spinars, magnetoids, etc.).

c) Massive black holes accreting gas or stars from their surroundings.

All these possibilities need fuller theoretical investigation. In particular, the most "conservative" theoretical ideas, involving frequent supernova outbursts in a dense gas cloud, may well prove versatile enough to explain everything yet observed. (See, for instance, Unno (1971).) Nor has anyone yet considered the detailed physics - including possible "magnetic flares", etc. - of accretion onto massive black holes.

Relevant observational questions undoubtedly include the following:

i) Can the variability be analysed into a succession of standard outbursts which involve the same "quantum" of energy in all objects? This might be expected in naive "building block" models involving successive supernova-type events - though even in these models the resultant luminosity may be sensitive to the density and pressure of the gas in the galactic nucleus, or to interactions between the debris from successive outbursts. If the variability resulted from (say) flares on a massive accretion disk, there would be <u>no</u> reason to expect the outbursts in different objects to involve similar amounts of energy. Two recent bits of evidence seem very hard to reconcile with naive "building block" models: the probable $\sim$ 1 day millimetre wavelength variability (Kellermann 1974) in the low-power source in the nucleus of Cen A which radiates $\sim 10^{49}$ ergs per day in all wavebands (mainly in X-rays); and the optical outburst in 3C 279 which lasted for several months and involved an energy output of $\sim 10^{54}$ ergs in the optical band alone (Eachus and Liller 1975).

ii) Is there a single centre of activity $\lesssim$ 1 pc in size, or do the outbursts "jitter" throughout a larger region?

iii) What is the shortest timescale on which substantial variations occur? (This is plainly also relevant to the nature of the radiation mechanism.)

If active galactic nuclei all involve qualitatively similar "non-thermal engines", we must then ask what secondary factors determine the particular mode (e.g. double radio source, Seyfert nucleus, quasar, "Lacertid"......) in which the activity manifests

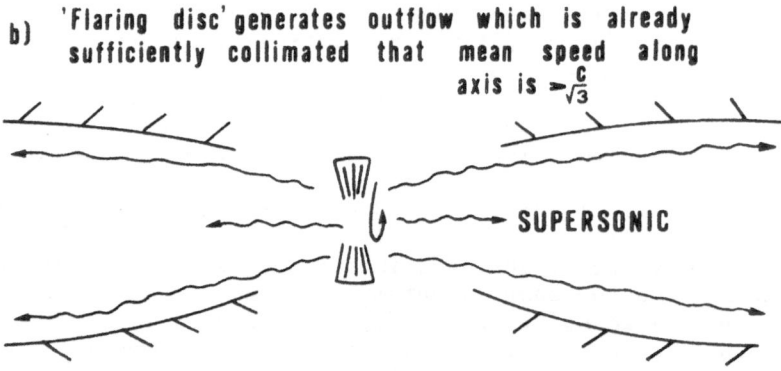

a)   Isotropic central source with collimation via nozzle mechanism

SHOCK

? Subsonic   Supersonic

pressure $\simeq P_o$

Shock occurs at radius R where $P/4 \pi R^2 c \; P_o$

b)   'Flaring disc' generates outflow which is already sufficiently collimated that mean speed along axis is $\sim \frac{c}{\sqrt{3}}$

SUPERSONIC

Fig. 5. Two possible origins for the hypothetical beams in double radio sources: a) the standard "twin exhaust" model – the central object produces an isotropic relativistic outflow, which becomes subsonic on passage through a shock (the variable compact radio sources being perhaps associated with the shock); b) if the central object were a flaring disk, the relativistic outflow could be generated with enough directionality to make the "nozzle" superfluous.

itself. These factors perhaps include:
    a) the power and duration of the central source;
    b) the density of the surrounding gas;
    c) the angular momentum of the surrounding gas.
    As already mentioned in 2.1, a) will be a major factor controlling radio source morphology. But the form of P(t) is also one of the factors determining the expected luminosity function for an ensemble of sources.
    The density of the surrounding gas will affect the strength of optical emission lines (though not perhaps in a straightforward way: the line strength may be limited by the amount of non-thermal

ultraviolet radiation able to photoionize the gas; and the gas may
develop a multi-phase filamentary structure, so that only a fraction of it emits optical lines).

Free-free absorption by dense gas surrounding a central non-
thermal source could in fact prevent any radio-band emission from
escaping. A spherically symmetrical cloud with radius R(pc) producing thermal bremsstrahlung emission of 10^{45} L_{45} erg sec^{-1} would
have unit free-free optical depth at a wavelength

$$\lambda \simeq 10^{-2} \ R(T/10^{40}K) \ L_{45}^{-1/2} \ cm.$$

This means that in quasars where compact radio structure is detected,
either the radio emission must predominately lie <u>outside</u> the thermal
gas; or else the gas must have a filamentary or disk structure, so
that it does not obscure the whole radio-emitting region.

Wall (1975) finds that the sources in the Parkes 2700 MHz
survey that have flat spectra or low frequency cutoffs are nearly
all quasars. This suggests that a compact radio source is always
associated with powerful non-thermal optical emission. The existence of radio-quiet quasars might seem to suggest that the converse
statement (i.e. non-thermal optical emission implies radio emission)
must be false: but perhaps the "radio quiet" quasars are simply
those where free-free absorption along our line of sight <u>is</u>
important.

The magnitude and orientation of the gas cloud's angular
momentum controls the directivity of any relativistic plasma outflow. This is clearly relevant to the origin of "beams". Moreover, the multiple absorption lines in quasar spectra may be caused
by filaments or sheets of thermal gas accelerated by the outflowing
plasma. The relationship of the so-called "Lacertids" to ordinary
quasars is of particular interest. These may be objects where
there is relatively little thermal gas: alternatively, maybe they
<u>do</u> contain gas, but the non-thermal ultraviolet continuum is too
weak to excite detectable emission lines.

Answers to these further questions would therefore be interesting:
    i) Are there radio-quiet "Lacertids"?
  ii) Are any "Lacertids" double radio sources?
 iii) Is double radio structure anti-correlated with the occurrence
of <u>multiple absorption</u> lines?

ACKNOWLEDGEMENT

I am grateful to many colleagues, particularly Dr. Roger Blandford,
for stimulating discussions on radio sources and related topics.

REFERENCES

Blandford, R.D. and McKee, C., 1975, Phys. Fluids, in press.
Blandford, R.D. and Rees, M.J., 1974, Mon. Not. Roy. Astron. Soc.
169, 395.
Blandford, R.D. and Rees, M.J., 1976, Rev. Mod. Phys., in press.
Colgate, S.A., Colvin, J.D. and Petschek, A.G., 1975, Ap. J. Lett.
197, L105.
DeYoung, D.S. and Axford, W.I., 1967, Nature 216, 129.
Eachus, L.J. and Liller, W., 1975, Ap. J. Lett. 200, L61.
Fomalont, E.B. and Miley, G.K., 1975, Nature 257, 99.
Gull, S.F. and Northover, K.J.E., 1973, Nature 224, 80.
Hoyle, F. and Fowler, W.A., 1963, Mon. Not. Roy. Astron. Soc.
125, 169.
Katz, J.I., 1975, Ap. J., in press.
Kellermann, K.I., 1974, Ap. J. Lett. 194, L135.
Kellermann, K.I., Clark, B.G., Niell, A.E. and Shaffer, D.B., 1975,
Ap. J. Lett. 197, L113.
Kompaneets, A.S., 1960, Sov. Phys. Doklady 5, 46.
Longair, M.S., Ryle, M. and Scheuer, P.A.G., 1973, Mon. Not. Roy.
Astron. Soc. 164, 243.
Lovelace, R.V.E. and Backer, D.C., 1972, Astrophys. Lett. 11, 135.
Mathews, W.G., 1971, Ap. J. 165, 147.
Ozernoi, L.M. and Ulanovsky, L.E., 1974, Sov. Astron. 18, 6.
Rees, M.J., 1971, Nature 229, 512.
Rees, M.J., 1975, Structure and Evolution of Galaxies, ed. G. Setti,
D. Reidel Publishing Co., Dordrecht, Nederland, p. 285.
Salvati, M. and Vitello, P., 1975, preprint.
Saslaw, W.C., 1974, Proc. IAU Symposium 58, ed. J.R. Shakeshaft,
D. Reidel Publishing Co., Dordrecht, Nederland, p. 305.
Scheuer, P.A.G., 1974, Mon. Not. Roy. Astron. Soc. 166, 513.
Spiegel, E.A., 1975, private communication.
Turland, B.D., 1975, Mon. Not. Roy. Astron. Soc. 170, 281.
Unno, W., 1971, Publ. Astron. Soc. Japan 23, 123.
Wall, J.V., 1975, Observatory, in press.

SOME PROBLEMS ASSOCIATED WITH NON-THERMAL RADIO SOURCES

G.R. Burbidge

Department of Physics, University of California-San
Diego, La Jolla, California USA

In these lectures I shall summarize work in three different areas.
They are:
  1. The optical identifications of non-thermal radio sources,
  2. Optical ejecta from radio sources, and
  3. The physics of the extragalactic compact non-thermal sources.

1. IDENTIFICATIONS OF RADIO SOURCES

It is now more than 20 years since the first non-thermal radio
sources were identified with optical objects. The present
situation is that many thousands of sources have been catalogued
in the radio surveys, and optical identifications have been tenta-
tively made for something of the order of a thousand sources.
Apart from the comparatively small number of sources which have
been identified with galactic supernova remnants (and the comparatively small number of thermal sources associated with gaseous
nebulae), the bulk of the sources have been identified with genuine
optical galaxies, N-systems, quasi-stellar objects and BL Lac
objects. In addition to this, a comparatively small number of
weak sources have been found to be associated with galactic stars.
These are discussed by Hjellming in lectures at this school. While
I shall not discuss them, I would like to stress the point that no
attempt has been made to identify radio sources with galactic stars
using the technique that is used to identify all of the extragalactic
sources. Instead, the stars which have been identified as radio
sources have been found simply by pointing the radio telescopes at
a chosen group of stars. I shall return to this point later.
  Spectroscopic studies have now been made and redshifts are

G. Setti (ed.), The Physics of Non-Thermal Radio Sources, 121–129. All Rights Reserved.
Copyright © 1976 by D. Reidel Publishing Company, Dordrecht-Holland.

available for about 500 QSOs and about 300 N-systems and radio
galaxies. Since a fraction, perhaps 10% - 20% of the QSOs are
not associated with radio sources, the total number of optical
objects associated with radio sources which have been identified
and for which redshifts have been obtained is therefore probably
somewhat more than 700. Here I am talking of "powerful" radio
sources and do not include normal spiral galaxies, or ellipticals
with weak radio emission from their nuclei.

At this point it is reasonable to ask whether there are any
misidentifications among these several hundred sources. To consider this question I briefly review the history of the identification
program.

Initially the objects identified on the basis of very crude
radio positions were galaxies with morphological pecularities.
They were Centaurus A(NGC 5128), Perseus A (NGC 1275), Virgo A
(M87) and Cygnus A which, although it was faint (17^m), showed both
morphological pecularities and had a unique optical spectrum.

In the succeeding years more identifications were made, and
the particularly important papers were those by Maltby, Matthews
and Moffet (1963) in which 24 more bright sources were identified
with galaxies, and the work by Matthews, Morgan and Schmidt (1964)
containing 52 more identifications. The early conclusions of
Bolton and this work led to the general belief, prevalent from
the early 1960's, that strong extragalactic radio sources are
identified with intrinsically very bright elliptical galaxies (with
or without emission lines in their spectra) and N-systems. The
morphological classification of D-systems and N-systems was due to
Morgan.

In the early 1960's the first identifications of quasi-stellar
objects were made. Since then the only new class of optical objects
that has been identified with radio sources is the class of lineless,
rapidly variable optical objects known as BL Lac objects.

An interesting question that should now be asked is: How many
misidentifications are there now in the literature? A related
question is: Are there other classes of optical objects still to
be identified as radio sources? While in the early days the radio
positions were only measured to precisions of many minutes of arc
or worse, the situation has steadily improved so that positions are
now often determined to accuracies of 10" or better. Thus it is
claimed that the chances of making a misidentification have continuously diminished.

However, it should be remembered that the majority of the
powerful extragalactic sources have large linear sizes. This means
that even if they are a long way away they have quite sizeable
angular dimensions (at a cosmological redshift z = 0.2, a source
with a "typical" linear size of 200 kpc has an angular size of
$\sim$ 30"). Since the optical source need not be at the centroid of
the radio source, this means that a considerable degree of uncertainty must remain in the identification based on position alone.
What other criteria are used? In the case of QSOs their apparent

rarity in the sky (using the statistics of one 6° x 6° field from
the work of Luyten and Sandage (1967)) makes if fairly safe for
us to assume that the identification is good even in cases in which
there is a considerable displacement between the position of the
QSO and the centroid of the radio source. It turns out that many
of the QSOs are sources of highly compact radio sources for which
very accurate radio positions can be determined. In these cases
the work of Sandage, Kristian and Wade (1970) has shown that in
most cases there is very good positional agreement – to a second
of arc or better – between radio and optical positions. There
are some ambiguous cases, e.g. 3C 94, where the displacement between the radio position and the QSO with which it has been identified is very large $\sim 40''$. Whether the identification should be
accepted has then been a matter of debate.

For sources which are not identified with QSOs what criteria
apart from the positions can be used to support or cast doubt on
the identification. There are two other properties of the optical
objects which have been used.

The first is the optical spectrum. We find that in all cases
but one, the exception being 3C 371, the N-systems which are identified on the basis of position as radio sources have strong broad
emission-line spectra. Since we suppose that such classes of
objects are rare (though no survey of N-systems which are not radio
sources has ever been made), the existence of a strong emission-
line spectrum is taken to confirm the correctness of the identification. For sources which are identified with bright elliptical
galaxies the situation is less clear. Many of these have normal
absorption-line spectra with weak emission (O II 3727, N II 6583,
Hα) sometimes present, but spectroscopically they are indistinguish-
able from the normal run of elliptical galaxies. It is often stated
that the radio ellipticals are predominately cD systems (in the
morphological classification). This may be true, but many are not,
and it is impossible to classify the many faint galaxies (> 18)
which are now being identified as radio sources. Thus, spectroscopic and morphological criteria frequently cannot remove the
ambiguity associated with positional identifications. Another
approach that has been attempted is to look for compact radio
sources on the optical objects which have been identified as the
sources of the extended radio sources. Apparently, compact components are detected in 30% to 50% of the sources investigated in
this way. This then will go some way to resolve the ambiguity.

In view of the difficulty associated with the positional
identifications and the faintness of the optical objects which are
being identified, it is hard to avoid the conclusion that a considerable number of misidentifications must already be present in
the literature. Some possibilities are:
  (i) that some radio sources identified with faint elliptical
      galaxies are really even further away and originate in
      galaxies which lie below the plate limit,
  (ii) that some sources originate in optical objects which are

physically associated with the galaxies which are called the identifications, but these optical objects are intrinsically faint and in general are below the plate limit. Such an idea would be compatible with the view that QSOs are "local" objects which have been ejected from galaxies. Cases like 3C 455 in which the galaxy was first identified, but then it turned out that the true identification was a QSO 23" away, would fit into this category. Misidentifications of this type would not affect calculations of the energetics and size of the source (because the distance is still correct), but they would affect our ideas about source generation and evolution.

(iii) Some sources may be galactic stars and not extragalactic objects at all. This possibility has not been considered seriously in recent years, but it should be looked into again. Radio stars clearly do exist, and formally the positional identification results do suggest that some galactic stars are the most likely identifications. In a considerable number of cases, the Lick group, the Texas group and others have found that many candidate QSOs turn out to be galactic stars. What has happened in general in such cases is that the observers, having found that the candidate QSO is a star with lines at rest wavelength, have dismissed the identifications out of hand.

This question should now be re-examined with care.

In concluding this discussion of the identification problems, I would like to discuss briefly the identifications and their implications for the 3CR radio catalogue. Out of slightly more than 300 sources, about 50 are QSOs and about 250 are not. Of these 250, 140 have been identified with galaxies, and about 100 of these have now had their redshifts measured. These galaxies fit quite well on the Hubble line. The mean redshift for these galaxies, $\bar{z} \simeq 0.14$. If we include the 40 for which we do not have redshifts, and assume that they have the mean intrinsic brightness of the bright radio galaxies for which redshifts are available, the mean redshift for the 140 identified galaxies goes up to $\bar{z} \simeq 0.19$.

How should we interpret the slope of the log N - log S curve for these sources? It is well known that the slope for all of these sources is about -1.8, and even after subtracting the 50 QSOs the slope remains the same. However, the log N - log S slope for the 140 sources identified with galaxies is slightly less (numerically) than -1.5. Thus, it is the remaining 110 sources which remain unidentified which give rise to the steep slope. It is commonly argued that the steep slope is caused by evolution in the populations of radio sources at epochs corresponding to redshifts in the range $z = 1$ to $z = 3$. If this is true, we can make a specific calculation to see whether it would be possible to identify and measure the optical objects at all. We know that they are not QSOs - at least not QSOs in the usual range of apparent magnitudes. If they are bright ellipticals, they will all be fainter than about 23^m and will not be measurable using ground-based telescopes. Thus, it will be impossible to confirm the evolutionary hypothesis by direct observation in the foreseeable future.

This chain of argument leads us to consider the possibility that already a considerable number of the faint galaxies in the 100 or 140, which we have treated as good identifications, are in fact misidentifications. Sources identified with many of the 18^m and 19^m galaxies may instead be really associated with:

(a) elliptical galaxies at much greater redshifts so that they are below the plate limits presently used, or

(b) they may be associated with a class of intrinsically fainter galaxies which are closer, or "local" QSOs which will fall below the plate limits at small redshifts.

2. OPTICAL EJECTA ASSOCIATED WITH RADIO SOURCES

We know that the violent activity in the nucleus of a galaxy gives rise to the generation of non-thermal sources both in situ, and on a very large scale outside the galaxies. It also gives rise to excitation of gas and also to the ejection of large amounts of gas. Is it possible that the violent events also give rise to the ejection of coherent objects? To many this might seem to be an unlikely possibility, but this is to be expected if QSOs are local objects ejected from galaxies, and also such objects may be responsible for the generation of extended radio sources. Rather than argue about the plausibility of the hypothesis, let us ask whether there is any observational evidence that such objects exist. There is some evidence as follows.

2.1 NGC 5128 (Centaurus A)

Many years ago Johnson (1963) showed that faint optical extensions are present in the direction of· the rotation axis of NGC 5128. Recent optical studies using the new 4-meter telescope at Cerro Tololo by Blanco, Graham, Lasker and Osmer (1975) and also work with the British 48-inch Schmidt telescope at Siding Springs have shown that faint filamentary structure and large numbers of what appear to be blue stars make up part of these faint optical extensions, and the geometrical configuration suggests that this material has been ejected from the central region of NGC 5128. Whether diffuse material has been ejected from NGC 5128 and has then condensed into stars, or whether stars themselves or other condensed objects have been ejected directly from the nucleus of NGC 5128 is not clear. However, the blue stellar objects are faint ($\sim 18^m$). Thus, it is clear that, if similar objects are also present in other powerful radio sources – even those as close as NGC 1275 and Cygnus A – they would be undetectable.

2.2 NGC 1275 (Perseus A)

It is well known that there is evidence for violent activity
throughout the Perseus cluster. Whether it all originates in
NGC 1275 is not clear. At one time it was thought that NGC 1275
was the source of radio emission throughout the cluster (Ryle
and Windram 1968). However, the most recent studies of IC 310 and
NGC 1265 have led to the conclusion that they are radio galaxies
in their own right. At the same time the very peculiar geometry
and dynamics of the bright galaxies in the Perseus cluster still
allows the free thinkers among us to admit the possibility that
many of these galaxies have been ejected as massive objects from
NGC 1275.

2.3 BO924+30

The configuration of this radio source discussed by Ekers, Fanti,
Lari and Ulrich (1975) does suggest that massive objects have
been ejected. There are three compact radio sources very precisely
aligned with the nucleus of the elliptical which lies at the center
of the two extended radio lobes. One of those is identified with
a 21^m blue stellar object for which spectra, as yet, have not
been obtained.

2.4 3C 40

This radio source lies in a cluster and is associated with the
bright ellipticals NGC 545, 547, and 541. No certain identification
has ever been made. There are extended regions of faintly luminous
material in the cluster, and a strange object originally identified
by Minkowski. Recent studies by Simkin have shown that it is a
strong emission-line object containing many large metal-poor H II
regions. In no way can it be considered to be a normal galaxy.
We can speculate that it is a massive coherent object which has
been ejected from the galaxy which is the source of the radio
emission.

2.5 3C 303

This radio source has been identified with an elliptical galaxy
which has a redshift z = 0.14. Kronberg (1976) has identified a
compact radio source component adjacent to the elliptical which
is centered on a faint optical object. The radio source configuration suggests that the optical object is associated with the
elliptical. Spectroscopic studies of this object have been carried
out by M. Burbidge and Smith at the Lick Observatory. It is
clear already that this object is neither a galaxy nor a galactic
star. This result in itself is very interesting. Again we might
argue that it is a massive coherent object which has been ejected
from the radio galaxy.

These observations suggest that coherent objects, which in some cases are detectable, are ejected from galaxies, In general such objects have not been looked for. To find more it will be necessary to search systematically in the lobes of strong radio galaxies at fairly small redshifts.

3. COMPACT NON-THERMAL SOURCES

Highly compact radio and optical sources are found in the nuclei of radio galaxies and in QSOs. Many of their properties have already been discussed here by K. Kellermann and M. Rees. I will therefore only briefly discuss a few points. Calculations concerned with compact radio sources have been made by Jones and Burbidge (1973), Jones, O'Dell and Stein (1974a, b) and Burbidge, Jones and O'Dell (1974). The point of the calculations has been to work out in some detail the magnetic fields, particle energies and total energies required to explain a variety of compact components in radio sources. It is assumed that the emission is incoherent synchrotron radiation due to a cloud of relativistic electrons in a random magnetic field. Variability is attributed to expansion of the cloud.

These compact sources are clearly associated with the optical non-thermal compact components. While the radiating mechanism is also the synchrotron process, we cannot get as much information from the observations. We know that in general the compact sources of optical radiation in both QSOs and the nuclei of galaxies are variable in time. No angular size measurements, comparable to those which are made by VLBI techniques at radio frequencies, are available and all that we can say is that if the components are star-like, they have angular sizes less than about 0".5. Also we do not know very much about the shapes of the spectra except to say that they are roughly of inverse power-law form. Upper limits to the sizes can therefore be obtained only from the variability, and we can calculate the energetics assuming either that there is approximate equipartition between electron energy and magnetic energy, or that the Compton effect is negligible. Such calculations were made by Demoulin and Burbidge (1968).

The main conclusions that one can draw from the results of these calculations are:

(i) For the compact radio components, in all cases but NGC 1275, the energy in the particles vastly exceeds the energy in the magnetic field, and total energies per compact component range from about 10^{52} to 10^{58} ergs. The values near the upper end of this range are obtained for QSOs like CTA 102 and 3C 454.3 which have large redshifts and are assumed to be at the distances obtained from those redshifts. Along with these results we must suppose that these highly energetic sources expand at highly relativistic speeds. The recent results of VLBI studies of the changes of angular size in 3C 273 show that the so-called "Christmas tree" model cannot be invoked. Jones and I (1973) showed that highly relativistic expansion in the presence of even an extremely low density surrounding gas can only be explained if energies much greater than the high values quoted below are present. There is no way out of this difficulty except that of invoking the local theory of QSOs. If they are much closer than the distances obtained from the redshifts, the total energies are reduced, and highly relativistic expansion does not have to be invoked.

(ii) For the highly compact optical components the energetics appear at first sight to be more moderate, lying in the range of $10^{52} - 10^{54}$ ergs per compact component. However, this result is obtained largely because equipartition, or near equipartition, is assumed, and if, as seems likely, the real situation is that the particle energy vastly exceeds magnetic energy, then the total energetics will be very similar to those found for the radio components. Also, it appears likely that the problem of highly relativistic motion, if the objects are at cosmological distances, will also have to be faced. The very large light variations seen in objects like 3C 279 also indicate that the energetic situation has been underestimated in the early calculations.

There is every indication that the energy released in each compact component is released in a short period, perhaps in a few years, so that we are talking of energy release at a rate of $10^{48} - 10^{50}$ erg sec^{-1}, or about $10^{55} - 10^{57}$ ergs per year, only a small part of this coming out in radiation and the bulk of it being dissipated by the escape of particles. This almost certainly rules out models in which it is supposed that individual supernova outbursts arising from the evolution of normal massive stars are responsible. In my view, not only is it likely that we are dealing with "local" QSOs, but we are seeing the collapse, implosion and explosion of massive, $> 10^3$ $M_\odot$ objects, at a rate of about one a year when we see compact components appear and evolve. Thus, the ultimate energy source is likely to be a cluster of massive objects, embedded perhaps in a cluster of more normal stars. Either this or something radically different, in the sense of Ambartsumian or Hoyle and Narlikar, is probably required.

REFERENCES

Blanco, V.M., Graham, J.A., Lasker, B.M. and Osmer, P.S., 1975, Astrophys. J. Letters 198, L63.
Burbidge, G.R., Jones, T.W. and O'Dell, S.L., 1974, Astrophys. J. 193, 43.
Demoulin, M.-H. and Burbidge, G.R., 1968, Astrophys. J. 154, 3.

Ekers, R., Fanti, R., Lari, C. and Ulrich, M.-H., 1975, Nature 258, 584.
Johnson, H.M., 1963, Pub. NRAO 1, 251.
Jones, T.W. and Burbidge, G.R., 1973, Astrophys. J. 186, 791.
Jones, T.W., O'Dell, S.L. and Stein, W.A., 1974a, Astrophys. J. 188, 353.
Jones, T.W., O'Dell, S.L. and Stein, W.A., 1974b, Ibid. 192, 261.
Kronberg, P.P., 1976, Astrophys. J. 203, L47.
Maltby, P., Matthews, T.A. and Moffett, A.T., 1963, Astrophys. J. 137, 153.
Matthews, T.A., Morgan, W.W. and Schmidt, M., 1964, Astrophys. J. 140, 35.
Ryle, M. and Windram, M.D., 1968, Mon. Not. Roy. Astron. Soc. 138, 1.
Sandage, A. and Luyten, W.J., 1967, Astrophys. J. 148, 767.
Sandage, A., Kristian, J. and Wade, C.M., 1970, Astrophys. J. 162, 399.

ENERGETICS OF QUASARS AND RADIO GALAXIES

L. Woltjer

European Southern Observatory
c/o CERN, Geneva, Switzerland

1. THE ENERGY REQUIREMENT

Simple estimates suffice to indicate that in quasars and radio
galaxies very large amounts of non-thermal energy are released.
In the case of quasars where the spectral distribution and polarization indicate that most of the radiation is likely to be non-
thermal, we start from the observed flux F at radio, infrared,
optical and X-ray wavelengths. Assuming isotropic radiation and a
cosmological interpretation of the redshift z we then may obtain
the luminosity L. The assumption of isotropy may be tested: from
the flux of Ly-α protons, the number of ionizing protons emitted
by a quasar may be estimated; comparison with an extrapolation of
the observed continuum suggests that isotropy is - to within a
factor of three or so - a reasonable assumption. If the time τ
during which the quasar has the luminosity L can be obtained, Lτ
gives a minimum energy estimate. The quasar 3C273 has a jet more
than 10^5 light years long. Taking on this basis τ equal to a few
times 10^5 years, we would obtain L$\tau \simeq 10^{60}$ ergs. With $\tau \simeq 10^5$ years,
the total number of quasars in the universe is comparable with the
number of giant galaxies. However, the quasars have a very broad
luminosity function with most being 2 - 5 magnitudes less luminous
than objects like 3C273. A possible interpretation is that quasars
are only for a small part of their lifetimes as bright as 3C273
and that most of the time they are fainter. If so the energy estimate we obtained should be reduced by at least a factor of ten. Of
course, it is also possible that the quasar luminosity function
reflects intrinsic differences so that 3C273 would always have been
very bright and other quasars would spend their entire lifetimes at
lower luminosity. Moreover, the value of L includes only the

G. Setti (ed.), The Physics of Non-Thermal Radio Sources, 131–135. All Rights Reserved.
Copyright © 1976 by D. Reidel Publishing Company, Dordrecht-Holland.

non-thermal energy which is converted into photons and more energy might go into other channels. We conclude that the total non-thermal energy produced in 3C273 is unlikely to be less than 10^{59} ergs, but could be quite a bit larger.

A more direct estimate of the total relativistic particle energy may be obtained for radio galaxies and quasars with radio emission in an extended volume. If we have relativistic electrons with a spectrum $n(E) = KE^{-\gamma}$ between energies E_{min} and E_{max} then according to standard synchrotron radiation theory the luminosity $L(\nu)$ approximately satisfies (with B the magnetic field strength and ν the frequency)

$$L(\nu) \propto KB^{\frac{1}{2}(\gamma+1)}\nu^{-\frac{1}{2}(\gamma-1)}$$

between frequencies ν_{max} and ν_{min} related to E_{min} and E_{max} through a relation of the form $\nu_m \propto BE_m^2$. The total energy of the electrons is given by

$$\varepsilon_e = \frac{K}{2-\gamma}(\varepsilon_{max}^{2-\gamma} - \varepsilon_{min}^{2-\gamma}) \propto L(\nu_o)B^{-3/2}(\nu_{max}^{\frac{1}{2}(2-\gamma)} - \nu_{min}^{\frac{1}{2}(2-\gamma)}),$$

where ν_o is some reference frequency. Typically for extended radio sources $\gamma \simeq 2.6$. Since $\nu_{max} \gg \nu_{min}$ we have approximately $\varepsilon_e \propto L(\nu_o)B^{-3/2}\nu_{min}^{-0.3}$. For the total magnetic energy ε_m we have

$$\varepsilon_m = \frac{B^2}{8\pi} \times \text{Volume.}$$

We see that the sum $\varepsilon_e + \varepsilon_m$ has a minimum (as a function of the unknown B) approximately for $\varepsilon_e = \varepsilon_m$ that is for equipartition of energy between electrons and magnetic field. Of course, this does not constitute a physical basis for equipartition. Also to obtain the total particle energy we have to add the energy of the protons and nuclei ε_p. If we now assume

$$\varepsilon_e + \varepsilon_p = \varepsilon_m$$

uniform filling of the volume with electrons and magnetic fields

$$\nu_{min} = 10 \text{ MHz}$$

$$\varepsilon_e = \varepsilon_p \,,$$

we obtain values of ε_e up to $10^{60} - 10^{61}$ ergs with equipartition fields of $10^{-4} - 10^{-6}$ gauss in the extended components of radio galaxies and quasars.

Evidence about the validity of equipartition assumption is hard to come by. In the case of Cen A the X-ray observations provide an upper limit on ε_e (through the inverse Compton process) which indicates that ε_e is not much above the equipartition value. As Dr. Kellermann explained, in the compact radio components of radio galaxies and quasars it frequently seems that $\varepsilon_e \gg \varepsilon_m$, but since the inferred ratio $\varepsilon_e/\varepsilon_m$ is proportional to a very high power of the angular diameter, one may worry about the sensitivity of the result to the particular, uniformly filled, model chosen. Parenthetically we note that in the Crab Nebula we seem to have approximate equipartition.

The assumption $\varepsilon_e \simeq \varepsilon_p$ is also uncertain. In the Crab Nebula it seems clear that ε_p cannot be much larger than ε_e, which may be relevant if the same acceleration process is operating in the radio galaxies and quasars. On the other hand, in the galactic cosmic radiation we have $\varepsilon_p \simeq 100 \; \varepsilon_e$.

Various observations of compact sources which have been discussed here suggest that the energy ε_e is not generated all at once, but rather in bursts occurring in the active sources once every few years which have minimum energies of perhaps $10^{54} - 10^{55}$ ergs, produced in regions less than a parsec across.

2. MECHANISMS FOR THE ENERGY PRODUCTION

Various proposals have been made concerning the objects involved in the acceleration of the relativistic electrons. In the following table we indicate the type of objects, crude estimates of the relativistic energy yield in units of c^2 per gram, the fraction of the energy going into relativistic particles and magnetic fields and the ratio of the energy in electrons (mass m) to that in protons (mass M).

Object	Yield	Relativistic particle fraction	e/p
Supernova explosion	10^{-4}	10^{-4}	m/M
Pulsar	10^{-4}	1	$(\frac{m}{M})^{1/3}$ to 1
Stellar collisions	10^{-3}	10^{-3}	m/M
Rotators Accretion flow on compact object $\Big\}$ 10^{-1}		10^{-1}	as pulsars?

In the hydrodynamic explosion of a supernova only a small fraction of the available energy comes off in relativistic particles,
the rest appearing as subrelativistic high velocity gas. This
fraction depends rather sensitively on the density profile in the
envelope of the presupernova. Since all particles are accelerated to the same velocity the electron fraction is small, unless
turbulent interaction between the protons and the surrounding
interstellar medium leads to a redistribution of energy. Assuming that most of the energy of a supernova explosion and of the
resulting pulsar is ultimately deposited in the supernova remnant, typical total energies of not much more than 10^{50} ergs
are inferred, corresponding to yields of 10^{-4} Mc^2. However,
the efficiency of acceleration of electrons seems to be much
better for the pulsars.

Stellar collisions in dense galactic nuclei could also release a fair amount of energy, but the efficiency of relativistic
particle acceleration would be rather low. A somewhat better
result could be achieved by a collection of fast moving pulsars,
where the efficiency of conversion of kinetic energy into relativistic electron energy might come closer to unity. However,
it would seem optimistic to believe that a large fraction of all
stars in a galactic nucleus would become supernovae.

Probably the most efficient energy production may take place
in rapidly rotating centrifugally supported objects - either in
the form of large coherent rotating bodies or in that of a steady
accretion flow on to a massive black hole. The physics of the
acceleration process could perhaps be similar to that in pulsars -
although the absence of a solid body in the middle may lead to
differences which have not yet been studied.

The general evolutionary picture that one might have if
massive rotators are involved is as follows. At first one has a
galactic nucleus sufficiently dense for relaxation effects to
be important. Because the relaxation time is proportional to
V^3/n (V being the velocity of the stars and n their spatial
density) or according to the virial theorem to $M^{1/2}R^{3/2}$ (M being
the mass and R the radius of the nucleus) it follows that relaxation (which leads to contraction) is a self-accelerating
process. As the density of stars increases coalescence may take
place and a massive rotating object may be formed. Assuming
conservation of magnetic flux one easily derives from standard
pulsar theory that the angular velocity of the rotator will vary
with time as $(1 - t/\tau)^{-\frac{1}{2}}$ with τ a constant proportional to the
mass and inversely proportional to the square of the magnetic
flux and of the angular velocity at t = 0. With increasing angular velocity the energy release would increase and one would
expect the maximum to be reached when the rotation velocity is
of order c and the radius close to the Schwarzschild radius.

If as an example we obtain the energy of a quasar out of a 10^8 solar mass rotator, we would expect a rotation period of 10^4 seconds in that phase - close to the most rapid variations that appear to have been observed. If we were to assume that all quasars have identical initial parameters, but are observed at random moments in their history, the luminosity function n(L) would vary as $L^{-7/4}$. Hence a very broad luminosity function would be expected.

The main problem during the evolution of the rotators is how to avoid large deviations from axial symmetry - which might cause most of the energy to be radiated in gravitational waves. Fragmentation instabilities might also be important. They actually might break up a massive rotator into a number of smaller objects. In this case the evolution of individual fragments might possibly be related to the burst-like behavior of the compact radio sources.

REFERENCE

Woltjer, L., 1974, Astrophysics and Gravitation (16th Solvay Conference Proceedings) p. 429 and references given therein.

THE ACCELERATION OF PARTICLES TO HIGH ENERGY

E.N. Parker

Department of Astronomy and Astrophysics
University of Chicago, Chicago, Illinois

1. CLASSES OF ACCELERATION EFFECTS

The common occurrence, and often spectacular consequence, of fast
particles in active astrophysical bodies has attracted the attention
of physicists for more than four decades. The acceleration mechanisms, whatever they may be, are remarkably efficient, converting
a major fraction of the total energy into fast particles. A
variety of ideas have arisen, suggesting how and why fast particles
are generated in various circumstances. The principal limitation
on particle acceleration theories has been the realization that
the universe is not filled with a hard vacuum, but rather is per-
vaded everywhere by tenuous ionized gases quite able to short
circuit any large-scale electric fields that occur under ordinary
circumstances. A number of the early ideas on the acceleration
of cosmic rays have been discarded for this reason. Gone are the
ambitious betatron acceleration schemes of Swaan and of Riddiford
and Butler (1952).
    In the presence of a thermal gas with high electrical conductiv-
ity, the energy of a fast particle can be increased only by adiabatic
compression and/or by the Fermi mechanism. Both of these effects
can be important in special situations. Considerable effort has gone
into investigation of circumstances in which the effective electrical conductivity of the fluid may be reduced (because of high current
density in a tenuous gas), or in which strong small-scale plasma
turbulence may be present, giving direct acceleration of particles
over and above the large-scale Fermi acceleration. Altogether
there are now several theoretical possibilities for acceleration
of particles to relativistic energies, although, as we shall see,
the high acceleration efficiency suggested by observation has not

G. Setti (ed.), The Physics of Non-Thermal Radio Sources, 137–167. All Rights Reserved.
Copyright © 1976 by D. Reidel Publishing Company, Dordrecht-Holland.

been demonstrated by any theory of which we are aware. The basic
theoretical ideas can be grouped roughly into five groups:
   1. Hydromagnetic Fields (in fluids of high conductivity)
      a. Fermi acceleration
      b. Compression by convection and diffusion in quasi-
         static magnetic fields
      c. Neutral sheet reconnection
   2. Field in Reduced Conductivity
      a. Neutral sheet reconnection
      b. Potential double layer
   3. Plasma Turbulence
   4. Low Frequency Electromagnetic Waves
   5. Supernova Explosion.
      We shall consider each in turn, attempting to expound the
basic physical effects and limitations and indicating the degree
to which the effectiveness and efficiency of the mechanism has been
established on sound principles. It is an unfortunate aspect of
the subject that particle acceleration occurs only under violent
and turbulent conditions, conditions that are generally difficult
to treat quantitatively. For that reason we cannot expect an elegant and complete theory. But it should be possible with serious
work in many cases to form an opinion as to the efficacy of the
basic idea. Unfortunately the tendency in modern astrophysics is
to effervesce on exotic cosmic phenomena without attention to the
basic effects.
      The reader should be aware that our review will not provide
an adequate answer to the question of particle acceleration. The
limitation of our theoretical ideas is perhaps best illustrated by
the solar flare, which is known from observations to convert 10-50
percent of its total energy into fast particles (electrons > 10 keV,
protons > 100 keV). There is no compelling reason, of which we
are aware, to think that any of our present ideas can convert so
large a fraction of the energy into fast particles.
      We begin our brief presentation with a consideration of the
electromagnetic fields in a highly conducting fluid, and progress
from there to the more special circumstances.

2. PARTICLE ACCELERATION IN A CONDUCTING FLUID

The electromagnetic fields in a fluid with a high electrical conductivity σ, and a velocity $\underline{v}$ small compared to the speed of light,
are easily deduced from Maxwell's quotations and the condition that
the electric field $\underline{E}'$ in the local frame of the fluid must be very
small. The fields $\underline{E}'$ and $\underline{B}'$ in the frame of the fluid are related
to the fields $\underline{E}$ and $\underline{B}$ in the fixed frame by the Lorentz transforma-
tions

$$\underline{E}' = \underline{E} + \underline{v} \times \underline{B}/c \qquad\qquad \underline{B}' = \underline{B} - \underline{v} \times \underline{E}/c$$

neglecting terms $O(v^2/c^2)$. If $\underline{E}'$ is small as a consequence of the high electrical conductivity, then

$$\underline{E} = -v \times \underline{B}/c \tag{1}$$

and, neglecting terms $O(v^2/c^2)$ again,

$$\underline{B}' = \underline{B}.$$

Now if the characteristic scale of the field is L, so that the characteristic time is $\tau = L/v$, it follows that $\partial E/\partial t$ is, in order of magnitude, equal to $vB/c\tau = v^2B/cL$. But Maxwell's equation is (in esu)

$$4\pi\underline{j} + \partial\underline{E}/\partial t = c\nabla \times \underline{B}$$

and the right hand side is of the order of cB/L. Thus $\partial\underline{E}/\partial t$ is smaller by a factor $O(v^2/c^2)$ and is to be neglected,

$$4\pi\underline{j} = c\nabla \times \underline{B} \tag{2}$$

Thus the current density $\underline{j}$ is of the order of $cB/4\pi L$, and from Ohm's law,

$$\underline{j} = \sigma\underline{E}'$$
$$= \sigma(\underline{E} + v \times B/c) \tag{3}$$

we have

$$\underline{E} = -\underline{v} \times \underline{B}/c + \underline{j}/\sigma,$$
$$= -\underline{v} \times \underline{B}/c + (c/4\pi\sigma)\nabla \times \underline{B}. \tag{4}$$

Maxwell's induction equation,

$$\partial\underline{B}/\partial t = -c\nabla \times \underline{E}$$

becomes the familiar hydromagnetic equation

$$\partial\underline{B}/\partial t = \nabla \times (v \times \underline{B}) - \nabla \times (\eta\nabla \times \underline{B}) \tag{5}$$

where η is the resistive diffusion coefficient $c^2/4\pi\sigma$ (in esu). The second term on the right hand side of (5) represents diffusion resulting from the second term on the right hand side of (4). It is smaller in magnitude than the first term by the reciprocal of the magnetic Reynolds number, $R_m = Lv/\eta$. Thus, the electric field component perpendicular to the magnetic field is

$$\underline{E}_\perp = -(\underline{v} \times \underline{B}/c) \left[1 + O(1/R_m) \right] \qquad (6)$$

The component parallel is

$$\underline{E}_\parallel = \underline{j} / \sigma$$

$$= (\nabla \times \underline{B})_\parallel \, \eta/c \qquad (7)$$

and is smaller than $\underline{E}_\perp$, by a factor $O(1/R_m)$. Since R_m is very large for astrophysical systems, $\underline{E}_\parallel$ is expected to be very small. Neglecting $\underline{E}_\parallel$, then, the equation of motion of a freely moving fast particle of mass M, charge q, and velocity $\underline{w}$ (which we may as well take to be small compared to c for the present discussion) is

$$M \, d\underline{w}/dt = q(\underline{E} + \underline{w} \times \underline{B}/c)$$

$$= q(\underline{w} - \underline{v}) \times \underline{B}/c.$$

Forming the scalar produce of $\underline{w}$ with this equation yields the rate of change of the kinetic energy of the particle,

$$\frac{d}{dt} \tfrac{1}{2} Mw^2 = -(q/c)\underline{w} \cdot (\underline{v} \times \underline{B})$$

$$= \underline{v} \cdot (q\underline{w} \times \underline{B}/c).$$

The quantity in parentheses on the right hand side is just the Lorentz force $\underline{F} = q\underline{w} \times \underline{B}/c$ on the particle. Hence the particle energy increases only when the fluid velocity $\underline{v}$ pushes against the Lorentz force, as when the particle is reflected from a head-on collision with a moving magnetic field, or compressed by an increasing field. This is, of course, just the Fermi mechanism of adiabatic compression. It is the only acceleration mechanism available within a highly conducting fluid limited to low velocities.

3. ADIABATIC COMPRESSION

The adiabatic compression of free particles in a magnetic field is thermodynamically no different from the compression in the absence of a field, of course. Consider the pressure exerted in a given direction by N particles per cm^3 each of mass M with a mean square velocity u^2 in that direction. The pressure is $p = NMu^2$, and is independent of the particle motions perpendicular to the direction. Hence the first law of thermodynamics for one dimensional adiabatic variations of a volume V in the chosen direction $(dU + pdV = 0)$ become

$$d(V\tfrac{1}{2}NMu^2) + NMu^2 dV = 0.$$

Integration of this equation yields

Nu^2V^3 = constant

Since conservation of particles requires NV = constant, this
becomes uV = constant. But for one dimensional compression, V is
simply proportional to the dimension ℓ. Hence $u\ell$ = constant.
This applies to compression in the absence of a magnetic field,
and to compression parallel to the magnetic field. For compression
perpendicular to a magnetic field, we must remember that the particle motions in the two directions are strongly coupled together
by the gyrations of the particles around the magnetic field with
the cyclotron frequency. Thus in terms of the mean square particle
velocity in either <u>one</u> of the directions perpendicular to the field,
the kinetic energy density is NMu^2 for the two directions. Thus

$$d(NMu^2V) + NMu^2dV = 0,$$

and

$$Nu^2V^2 = \text{constant},$$

so that

$$u^2V = \text{constant}.$$

But now $V \propto \ell^2$ and conservation of magnetic flux requires $B\ell^2$ =
= constant. Hence

$$u^2/B = \text{constant}.$$

In more conventional notation one usually writes $w_\perp$ for the rms
particle velocity in the two dimensions perpendicular to B, so
that $w_\perp^2 = 2u^2$. Then for adiabatic compression of a magnetic
field B (without compression in the direction parallel to B) one
writes

$$B/N = \text{constant}, \quad \tfrac{1}{2} Mw_\perp^2 /N = \text{constant}$$

or

$$\mu \equiv \tfrac{1}{2} Mw_\perp^2 /B = \text{constant}$$

Charged particles flitting back and forth along a field with
speed $w_{\parallel}$ between approaching magnetic mirrors are also compressed
adiabatically, with the result that

$$N\ell = \text{constant}, \quad w_{\parallel}\ell = \text{constant},$$

where ℓ is the separation of mirrors.

If the region undergoes simultaneous transverse and longitudi-

nal compression, these results are easily combined, beginning
with the basic conservation relations

   $A\ell N$ = constant, AB = constant,

where A is the cross sectional area. Then again $w_\perp^2/B$ = constant
and $w_\parallel/\ell$ = constant, so that $w_\parallel B/N$ = constant. The variation of
the particle velocities follows in terms of their number density
N and the field strength B. The particle pressures $p_\perp = \frac{1}{2} NMw^2$
and $p_\parallel = NMw^2$ vary as $p_\perp/NB$ = constant and $p_\parallel B^2/N^3$ = constant.
      In a strong dipole field such as that of Earth, there are
possibilities for particle acceleration as a consequence of convection and diffusion in the magnetic field. Of basic importance
to the Van Allen belt of Earth, and the very intense radiation
belt of Jupiter, is the fact, already noted, that the magnetic
moment μ of a particle trapped in a field B is conserved for slow
variations of B, while the particle migrates deep into the field.
Thus, a proton may be injected with modest energy of 500 eV into
the geomagnetic field at the magnetopause, where $B = B_1 = 5 \times 10^{-4}$
fauss. It drifts around the distorted dipole field along a line
of constant B. The continual agitation of the field causes the
contours of B to fluctuate and the particle to diffuse radially,
with the result that it may move in toward earth. Under steady
conditions the radial diffusion (Kellogg, 1959; Parker, 1960;
Davis and Chang, 1962) produces a particle density distribution
that may increase inward toward a stronger field. The interesting
point is that those particles which eventually find themselves
close to Earth, in a strong field, B_2, have a kinetic energy larger
than their initial injection energy by a factor B_2/B_1. At a radial
distance of $2R_E$ ($R_E = 6.4 \times 10^3$ km = radius of Earth), we have
$B_2 = 4 \times 10^{-2}$ gauss, and the energy has increased by a factor of
10^2. An initial 0.5 keV becomes 50 keV. The circular motion
$w_\perp$ has been compressed, conserving angular momentum, and heating
the particle accordingly.
      At the same time that $w_\perp$ is increasing, hydromagnetic waves
may cause the mirror points of the particle to move up and down,
increasing $w_\parallel$ in the manner described in (Parker, 1961).

4. FERMI ACCELERATION

To treat the Fermi mechanism in its simplest form (Fermi, 1949) we
imagine that the particle is repeatedly reflected from randomly
moving magnetic inhomogeneities, each with a velocity v. If the
mean free path between collisions is L, then in a simple one dimensional system, the rate of head-on and overtaking collisions
is $(w + v)/L$ and $(w - v)/L$, respectively, given that the total
rate is w/L. The particle is reflected in the frame of reference
of the moving inhomogeneity, so that, following the collision, the
velocity in the fixed frame is $w + 2v$ for a head-on collision and

w – 2v for an overtaking collision. The gain in kinetic energy, $T = \frac{1}{2} Mw^2$, is $2Mv(w + v)$ and $2Mv(w - v)$ respectively. The mean rate of change of kinetic energy is, then,

$$\frac{dT}{dt} = \frac{w + v}{2L} \ 2Mv(w + v) - \frac{w - v}{2L} \ 2Mv(w - v)$$

$$= 4Mv^2w/L,$$

or

$$\frac{1}{T} \frac{dT}{dt} = 8 \ \frac{w}{L} \ \frac{v^2}{w^2} \qquad (8)$$

where w/L is the collision rate. Thus the mean fractional change in kinetic energy per collision is of the order of v^2/w^2. It is a net gain because head-on collisions are more probable, and involve a bigger energy change, than overtaking collisions. The numerical coefficient 8 is not to be taken seriously in a three dimensional universe. A better estimate might be 2. Had we carried out the calculation for relativistic velocity $w \simeq c$, we would have obtained essentially the same result, neglecting terms $O(v/c)$ compared to one.

If we restrict attention to the simple extreme relativistic case that $w \simeq c$ and suppose that the mean free path L between collisions is independent of T, then,

$$\frac{dT}{dt} = \alpha T \qquad (9)$$

where

$$\alpha = s(c/L) \ v^2/c^2 \ , \qquad (10)$$

(with s of the order of unity) is a constant, independent of T. Denote by F(T)dT the number of particles in the energy interval (T, T+dT) in a large statistically homogeneous region of "turbulent" magnetic field. The conservation of particles leads to

$$\frac{\partial F}{\partial t} + \frac{\partial}{\partial T} \left(\frac{dT}{dt} \ F \right) = 0.$$

If we imagine that particles are introduced at a low energy T_o and escape when they reach T_1, then under steady conditions we have

$$F(T) = \frac{C}{dT/dt} = \frac{C}{\alpha T} \qquad (T_o < T < T_1)$$

where C is a constant. If the rate of injection at T_o is N_o, then

$$N_o = F(T_o)(dT/dt)_o$$

$$= F(T_o) \ \alpha \ T_o$$

and $C = N_O$ so that

$$F(T) = (N_O/\alpha T_O)T_O/T. \tag{11}$$

The energy spectrum is just T^{-1}. Energy spectra are generally not this flat, usually $T^{-\nu}$ with $\nu > 2$.

Suppose that as a consequence of strong interactions with nuclei in the background plasma, or escape from the confining field, a particle has a mean life τ that is independent of T. The conservation of particles leads to[†]

$$\frac{\partial F}{\partial T} + \frac{\partial}{\partial T}(\alpha TF) + \frac{F}{\tau} = 0.$$

Steady conditions yield

$$F(T) = \frac{N_O}{\alpha T_O}\left(\frac{T_O}{T}\right)^{1+1/\alpha\tau} \tag{12}$$

The spectrum is steeper by the additional power $1/\alpha\tau$.

For cosmic rays $F \propto T^{-2.7}$ so that $\alpha\tau \simeq 0.6$. The cosmic rays filling the gaseous disk of the galaxy have life in the disk of about 3×10^6 years $= 10^{14}$ sec, based on a penetration through about 5 gm/cm^2 of material in interstellar space to account for the observed abundance of collision fragments such as H^2, He^3, Li, Be, B (Silberberg, 1966; Shapiro and Silberberg, 1970). It follows, then, that α must have a value of 0.6×10^{-14} if we are to account for the observed cosmic ray spectrum by Fermi acceleration in interstellar space. The observed interstellar gas motions are typically $v = 10$ km/sec over scales of $L = 100$ pc $= 3 \times 10^{20}$ cm. We compute from (10) that $\alpha = 2 \times 10^{-19}$/sec, too small by a factor of 10^5. One can, of course, use very small values of L and large values of v (Morrison, Albert, and Rossi 1954) to increase the estimate of α.

Quite recently Scott and Chevalier (1975) and Chevalier (1975) have pointed out that the Fermi mechanism seems to work quite well in the high velocity, small scale turbulence in supernova remnants. They estimate turbulence velocities of $v = 6 \times 10^3$ km/sec and scales of $L = 4 \times 10^{16}$ cm, so that $\alpha = 6 \times 10^{-10}$. Thus if the mean life of an accelerated particle within the remnant is 30 years before escaping, we have $\alpha = 0.6$ and a spectrum like that observed for cosmic rays. They estimate that protons can be accelerated rapidly and efficiently up to 5×10^{15} eV in the Cas A supernova remnant, and that there are enough such supernova remnants continually created in the disk of the galaxy to produce and maintain the present cosmic ray intensity. The idea is interesting, certainly, and needs further exploration to understand, if possible, whether, and how, sufficient energy is fed into the turbulent magnetic fields in the supernova

---

[†]See also the point made by Davis (1956), that there need not be a net energy gain.

remnant to perform the colossal task that they attribute to it.

In view of the limitations of (10) being second order in v/w, Fermi (1954) suggested that there might be circumstances in which head-on collisions predominated. Then, in place of (8) we would have

$$\frac{1}{T}\frac{dT}{dt} = 4\frac{w}{L}\frac{v}{w} \tag{13}$$

so that $\alpha = 4(w/L)(v/w) = 1.3 \times 10^{-14}/\text{sec}$, which just nicely does the job for interstellar space. There are difficulties, of course. The energy density of the cosmic rays in the disk of the galaxy is approximately $1 \text{ eV/cm}^3 = 1.6 \times 10^{-12} \text{ ergs/cm}^3$, so that the suggested life of 10^{14} sec (Silberberg, 1966; Meyer, 1969; Shapiro and Silberberg, 1970) implies a production rate of $1.6 \times 10^{-26} \text{ ergs/cm}^3$ sec throughout the gaseous disk. This is somewhat more than the conventional estimates (of $0.1 - 0.5 \times 10^{-26} \text{ ergs/cm}^3$ sec) of the energy input to the turbulence of the interstellar gas (Parker, 1968, 1969; Silk, 1975) from hot luminous stars, but the present state of uncertainty does not permit a dogmatic conclusion in this respect. In any case, if we were to argue that the principal acceleration of cosmic rays occurs in interstellar space, then the Fermi mechanism must be very efficient to channel most of the available energy into the acceleration of cosmic rays, rather than losing it to radiation, etc., as is generally believed. It is curious, too, that the iron nuclei among the cosmic rays evidently must have a somewhat shorter life in interstellar space than protons because their collision cross section is so large. Yet their energy spectrum appears to be flatter (larger $\alpha\tau$), if anything, than the protons.

It is curious, too, that the radio spectra of other galaxies suggest cosmic ray electron spectra similar to the spectrum in our own galaxy, implying that $\alpha\tau$ is universally a little less than one[†]. Yet there is no reason to expect the same v and L in other galaxies, some of which are much more active and bear no resemblance to our own. Clearly there is more to the problem than the simple ideas presented so far. Perhaps there are some dynamical effects in which the particles interact sufficiently strongly with the fluid and field motions so as to produce a universal particle turbulence spectrum. Syrovatskii has proposed hypothetical circumstances wherein such effects might occur, giving a universal spectrum.

Unfortunately the enormous theoretical difficulties in treating the interaction of particles and plasma turbulence has limited the work so far to weak turbulence (so that the quasi-linear approximation can be used) (see Sec. VII) and to interactions in a closed

---

[†]The fast particles produced in the geomagnetic field and in solar flares do not conform to the universal galactic spectra, being generally much steeper.

system so that the radiation fields can build up to equilibrium
with the particles and plasma turbulence. The ideas are fascinating
but do not yet seem to strike at the heart of the real problem of
copious production of fast particles.

To come back to the Fermi mechanism, consider a specific
case of (13). We suggested (Parker, 1958) that sharp crested
Alfven waves in the galactic field might give the desired prominence
of head-on collisions. Imagine a particle with pitch angle $\theta_o(t)$
in a field B_o trapped between two approaching Alfven waves with
maximum field density B_w (> B_o) (i.e., $\sin^2 \theta(o) > B_o/B_w$ as sketched
in Figure 1. The particle reflects back and forth between the two
waves. The kinetic energy of the motion perpendicular to the field
satisfies

$$\frac{1}{2} Mw_\perp^2/B = \text{constant} \tag{14}$$

and its value in the uniform field B_o between the waves does not
change $w_\perp(t) = w_\perp(0)$. The motion parallel satisfies

$$w_{\parallel}(t)\ell(t) = \text{constant} \tag{15}$$

where $\ell(t)$ is the separation of the two waves. The pitch angle
θ is given by $\tan\theta = w_\perp/w_{\parallel}$. In the increased field of one of the
waves the pitch angle $\theta(t)$ increases from its value $\theta_o(t)$ in the
uniform field B_o,

$$\frac{\sin^2\theta}{B} = \frac{\sin^2\theta_o}{B_o} \tag{16}$$

in the frame of the wave. The increase of w with successive
reflections (all of which are head-on collisions) decreases $\theta_o(t)$

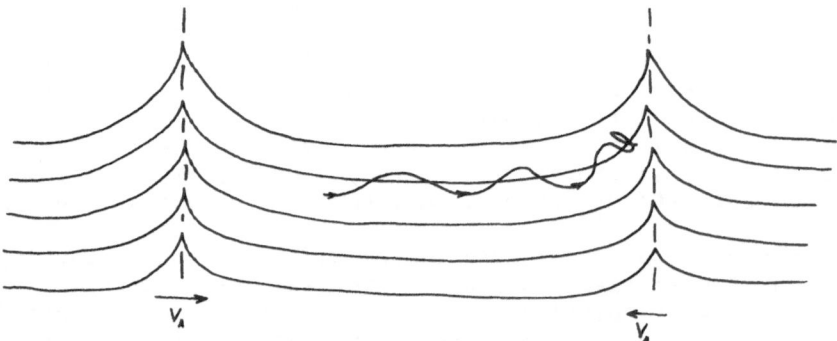

Fig. 1. A sketch of two Alfven waves with sharp crests approaching
each other along a uniform magnetic field B_o. The wavy line
indicates a particle trajectory in the field, showing reflection
in the increased field of the wave.

until finally $\theta(t)$ fails to reach $\pi/2$ by the time $B = B_w$, and the particle escapes by penetrating through the wave. This occurs when $\theta_0(t)$ has decreased to

$$\sin^2\theta_0(t) = B_0/B_w \qquad (17)$$

from its initial value $\theta_0(0)$. At this point

$$w_\parallel{}^2 = w_\perp{}^2\cot^2\theta_0(t),$$
$$= w_\perp{}^2 \, B_w/B_0 - 1,$$

and

$$w^2(t) = w_\parallel{}^2(t) + w_\perp{}^2,$$
$$= w_\perp{}^2(1 + \cot^2\theta_0),$$
$$= w_\perp{}^2 \, B_w/B_0.$$

If the initial pitch angle was $\theta_0(0)$ when the particle first entered the region between the approaching waves, then the initial velocity was

$$w^2(0) = w_\perp{}^2 \left[1 + \cot^2\theta_0(0) \right]$$

and there has been a gain in kinetic energy by a factor

$$\frac{w^2(t)}{w^2(0)} = \frac{1 + \cot^2\theta_0(t)}{1 + \cot^2\theta_0(0)} = \frac{B_w}{B_0} \, \sin^2\theta_0(0) \qquad (18)$$

by the time the particle escapes. If, for instance, the initial pitch angle were $\theta_0(0) = \pi/3$, and $B_w/B_0 = 2$, so that $\theta_0(t) = \pi/4$, then

$$w^2(t)/w^2(0) = 3/2$$

and the kinetic energy has increased by fifty percent. The increase occurs in a time ℓ/V_A, and so yields (13). A proper average over all angles can be carried out (Parker, 1958) confirming the approximate ratio of 3/2.

Now if there were nothing more to the story than this, the particles escaping from the wave trap would have a pitch angle too small ever to be caught by a similar wave again. The point is, then, that the sharp crest of the wave, or other small-scale inhomogeneities, is called upon to scatter the particles as they pass through, redistributing them to isotropy and setting them up for the next two approaching waves (Parker, 1958). If the sharp crest should fail to accomplish this, then there are relativistic plasma instabilities (Lerche, 1967) that feed on the anisotropy and quickly reduce it to low values.

Altogether, we can account for much of the acceleration of cosmic rays (once they are injected at energies of a few hundred MeV to avoid heavy ionization losses) by Fermi acceleration in interstellar space by using conventional numbers such as v = 10km/sec and ℓ = L = 100 pc. Of course, it is not clear that the interstellar motions represent a suitable array of Alfven waves of large amplitude propagating along the field. Nor is it clear that there is enough energy being pumped into the waves (10^{-26} erg/cm^3 sec) to sustain the cosmic ray acceleration. And if there is, we have yet to show that the acceleration of cosmic rays is the principle energy loss to the waves. For if there are competing forms of energy loss, the energy is wasted.

It has been suggested (Parker and Tidman, 1958; Parker, 1958) that the Fermi mechanism may produce superthermal ions in most agitated plasmas, such as the solar corona, the geomagnetic tail, and (Jokipii, 1968) in the solar wind.

5. NEUTRAL SHEET ACCELERATION

A particularly interesting mechanism for particle acceleration, suggested for the geomagnetic tail as well as solar flares, is the acceleration in neutral sheets, between two regions of oppositely directed magnetic field (Hoyle, 1949; Dungey, 1953). To illustrate this idea, consider a magnetic field B = e_z B(x) which is an odd function of x, vanishing on the yz plane (x =o) and having the same sign as x on either side. Suppose that B(x) approaches a uniform value ±B_o for large x. Hydrostatic equilibrium in the x-direction requires that the sum of the gas pressure and the field pressure be uniform,

$$p(x) + B^2(x)/8\pi = p_o \tag{19}$$

where p_o is a constant. Since B(0) = 0, it follows that p(x) is a maximum at x = 0, with a value p_o. Elsewhere the gas pressure is less, as sketched in Figure 2. But in that case there is nothing to keep the high pressure gas in place at x = 0. Its pressure is above the ambient value, and either it flows away along the lines of force, escaping out the edge of the field, or it escapes in blobs out through B due to large-scale hydromagnetic exchange instabilities. Special circumstances would be required to prevent its escape. Now if the gas escapes, it lets the two regions of opposite field come closer together, but of course (19) is still satisfied, so the escape of gas continues and the two regions continue to approach more closely, the field gradient between becoming steeper and steeper. If the transition layer between the two opposite fields has a characteristic thickness 2ℓ, shown in Figure 2, then the current density follows from (2) as

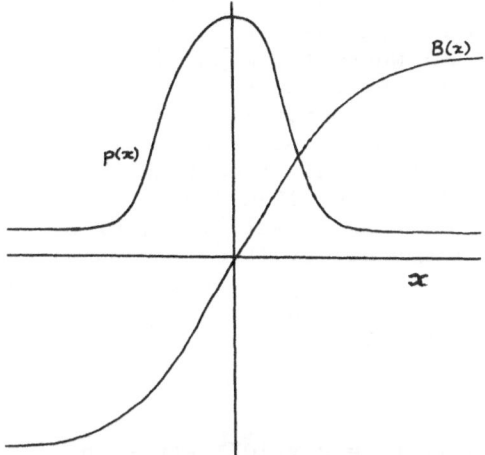

Fig. 2. A sketch of the variation of gas pressure p(x) and magnetic field strength B(x) across a neutral sheet.

$$j = - \frac{c}{4\pi} \frac{dB}{dx} = 0 \left(\frac{cB}{4\pi \ell} \right) \qquad (20)$$

in the y-direction. As ℓ diminishes, the current density becomes
larger. It follows from (4) that there is an electric field in
the y-direction given by

$$E = - \frac{\eta}{c} \frac{dB}{dx} + \frac{v(x)B(x)}{c} \qquad (21)$$

where v(x) is the x-component of the fluid velocity. If we imagine
steady conditions, wherein field and gas are convected in from each
side with a velocity v(x) as rapidly as they are expelled and dissipated in the neighborhood of x = 0, then $\partial \underline{B}/\partial t = 0$ and hence
$\nabla \times \underline{E} = 0$. It follows that $\underline{E}$ must be independent of both x and z.
Since all quantities are taken to be independent of y anyway, it
follows that $\underline{E}$ = constant, with a value (y-direction)

$$E = - \frac{\eta}{c} \left(\frac{dB}{dx} \right)_{0} \qquad (22)$$

where the subscript denotes the value at x = 0. Hence at large
$x (>> 1)$ where $B \sim B_{0}$ we have

$$E = vB_{0}/c$$

or

$$v = cE/B_{0}$$
$$= - \frac{\eta}{c} \left(\frac{dB}{dx} \right)_{0} .$$

It has been suggested that the electric field E in the y-direction
may accelerate particles to very high energies. But this requires
that E must be very strong. In the presence of a highly conducting
fluid this is restricted to a single circumstance in which the

opposite fields are in contact only over a narrow region, as sketched
in Figure 3 (Petschek's (1964) mechanism) where the magnetic lines
of force are shown by the solid lines and the fluid velocity by
the dashed lines. Our equations apply in an approximate way to
conditions along the x-axis (z = 0). With the short region of
contact, of length L, the gas squeezes out rapidly from between the
fields. Calculations (Petschek, 1964; Sonnerup, 1970; Yeh and
Axford, 1970; Fukao and Tsuda, 1973a, b; Yeh and Dryer, 1973; Priest,
1973; Vasyliunas, 1975) indicate that L may adjust itself to so
small a value that the rate at which the fields move together is
$v = \epsilon V_A$ where V_A is the Alfven speed $B_0/(4\pi\rho)^{1/2}$ and ϵ is a number
of the order of 10^{-1} or greater. In this case

$$E = \epsilon B_0 V_A/c$$

and can be quite large. For instance, if $B_0 = 2 \times 10^{-4}$ gauss in
the geomagnetic tail where there is one hydrogen atom per cm^3,
then $V_A = 400$ km/sec and $E \simeq 3 \times 10^{-8}$ statvolts/cm = 10^{-6} volts/cm.
The potential difference across the 10^{10} cm width of the tail would
then be 10^4 volts. Much larger voltages can be calculated for the
solar flare (B = 10^3 gauss, N = $10^{11}/cm^3$, $V_A = 7 \times 10^8$ cm/sec,
L = 10^9 cm yield 6×10^{11} volts). Speiser (1965, 1967) and

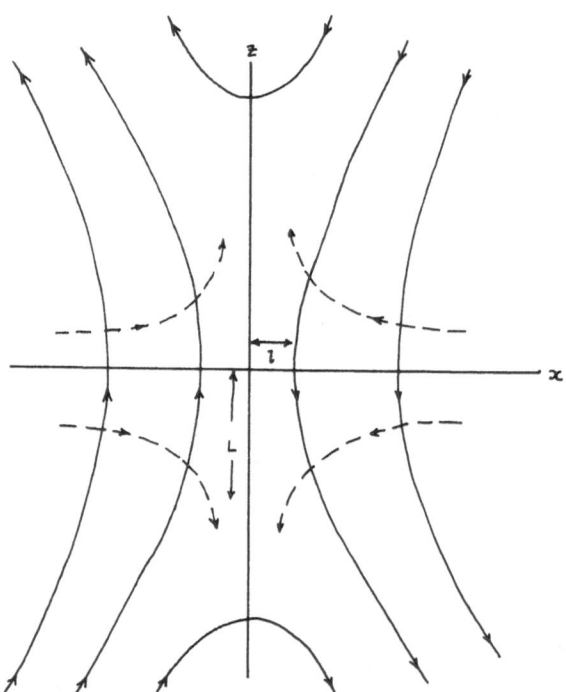

Fig. 3. A sketch of the magnetic lines of force (solid lines) and
the fluid motion (dashed lines) where two opposite fields meet over
a narrow region (Petschek's mechanism) and reconnect.

Sonnerup (1971) have calculated particle orbits in a neutral sheet, to show along what trajectories particles are accelerated. Low (1974) has worked out the particle acceleration for specific field models. Particles of large energy result, as indicated above, but unfortunately, if the geometry must be restricted to the narrow region of contact sketched in Figure 3, only a <u>small</u> fraction of the energy is available to accelerate particles. Most of it goes into bulk acceleration of the streaming fluid. Only in the near neighborhood of the small dissipation region $2\ell \times 2L$ where $\underline{B} \simeq 0$ is it possible for a free charge to pick up energy from E. Elsewhere the electric field experienced by the particles is essentially zero because all particles move with the electric drift velocity $\underline{u} = c\underline{E} \times \underline{B}/B^2$, in which frame $\underline{E}' = 0$. Therefore the idea has its limitations and appears to be unsuitable for a flare, where it is estimated that as much as half the energy may go into fast particles.

Various instabilities have been appealed to, including the tearing mode illustrated in Figure 4, which is, in fact, merely a small scale repetition of Figure 3. But the authors of ideas that invoke Figure 4 fail to state how the fluid escapes from the region once the rapid dissipation and reconnection, indicated in the thin transition layer, is finished and the opposite fields are separated by a layer of fluid. If the fluid remains between the opposite fields, the reaction will stop. Large scale exchange instabilities may perhaps remove it (Parker, 1973) at a suitable rate, comparable to the Alfven speed. One may imagine that if the opposite fields meet over a broad front, as in Figure 4, there is more opportunity for the acceleration of charged particles, and hence a higher fraction of the initial magnetic energy can be converted into fast particles.

Further ideas on particle acceleration appeal to special effects to reduce the electrical conductivity of the medium.

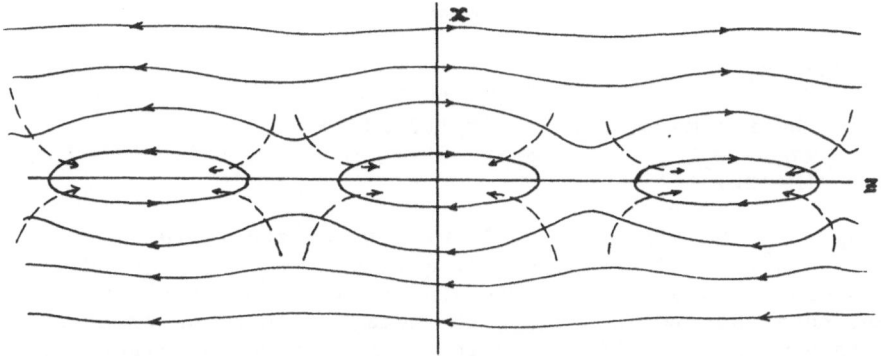

Fig. 4. A sketch of the magnetic lines of force (solid lines) and fluid motion (dashed lines) in the transition region between two opposite fields, depicting the tearing mode instability and the many local reconnections of the magnetic lines of force.

6. PARTICLE ACCELERATION WITH REDUCED CONDUCTIVITY

There are a number of ideas that have been originated to suggest
that the theoretical electrical conductivity,

$$\sigma = 10^7 \, T^{1/2} \text{ esu,} \tag{24}$$

valid for small current densities in an ionized gas, may be greatly
reduced at critical points in the system where the current densities
are high. Presumably the enhanced resistivity would increase the
dissipation and electric field in neutral sheets, as well as per-
mitting strong electric fields parallel to the magnetic field.
The possibility of enhanced resistivity opens up a whole new realm
of acceleration ideas.

For instance, if the current density parallel to $\underline{B}$ is so high
that the electron conduction velocity u exceeds the ion thermal
velocity $(3kT/M_i)^{1/2}$, the various micro instabilities (Alfven and
Carlquist, 1967; Carlquist, 1969; Hamburger and Friedman, 1968;
Kalinin et al. 1970; Smith and Priest, 1972; Papadopoulos and
Coffey, 1974) lead to plasma turbulence and to enormously enhanced
resistivities within which strong parallel electric fields may be
possible. If there are N_e electrons per cm^3, then $j = -N_e eu$ and

$$u = -j/N_e e.$$

In terms of the change B in the magnetic field over the scale ℓ,
this is

$$u = -cB/4\pi N_e \ell e. \tag{25}$$

It is clear that if N_e is small enough, u can be larger than the
ion thermal velocity. This requirement

$$u^2 > 3kTe/M_i \tag{26}$$

where M_i is the ion mass, can be simplified to

$$\ell w_p < c(M_i/M_e)^{1/2}$$

upon noting that the gas pressure $2N_e kT$ between the fields is
comparable to $B^2/8\pi$ and the plasma frequency is $\omega_p = (4\pi N_e e^2/M_e)^{1/2}$.

To understand under what circumstances we might reach this
condition, note that for a scale ℓ the normal rate of destruction
of field is $v = \eta/\ell$. So if, under the circumstances of Figure 3,
$v = \varepsilon V_A$, then

$$\ell = \eta/\varepsilon V_A.$$

The condition (26), with u given by (25), becomes

$B\sigma\epsilon > N_e ec$

upon writing $B^2/8\pi = 2N_e kT$ again. With $\epsilon = 0.1$ this is

$B\sigma/N_e \gtrsim 10^2$

or

$BT^{3/2}/N_e \gtrsim 10^{-5}$

with the aid of (24). The inequality is satisfied for strong fields, high temperature (i.e. high conductivity) and small density. In the solar chromosphere over an active region where B = 500 gauss, T = 2 x 10^4 °K and $N_e \simeq 10^{12}/cm^3$, we have $BT^{3/2}/N_e \simeq 10^{-3}$. The equality is easily satisfied. If the electrical conductivity should drop precipitously, then according to (22) the electric field could become quite large, and over a much broader front than indicated in Figure 3 (Coppi and Friedland, 1971). The work of Smith and Priest (1972) suggests, however, that most of the energy then goes into heating the background plasma, rather than into producing a much smaller number of very fast particles. So the idea, although attractive, is difficult to demonstrate.

Michel (1968, 1969) has proposed that the gas density is so low (10^{-5} protons/cm^3) in such places as the Crab Nebula that the disordered fields there produce large quantities of fast particles in the neutral sheets between regions of opposite field.

It has been suggested that within the magnetosphere of the Earth particles are accelerated by electric fields parallel to the magnetic field (Alfven and Carlquist, 1967; Evans, 1974; McIlwain, 1975), something that is absolutely forbidden in the presence of a very high electrical conducitivity. The ideas go beyond the enhanced resistivity, caused by microturbulence in regions of high current and low plasma density, to the potential double layer. The double layer is presumably a self consistent electrostatic plasma sheath in an otherwise highly conducting plasma, in which the current is limited by space charge effects, so that the effective conductivity is very low, i.e. the plasma density is very low compared to the current density. It has been suggested as a commonly occurring effect in the convecting magnetosphere of the Earth, and we shall present it here in that context.

First then, a few words on the convection, electric fields, and electric currents in the magnetosphere (Axford and Hines, 1961; Axford, 1964; Walbridge, 1967). We treat an idealized model of a vertical field extending up from a flat Earth.

Consider a uniform magnetic field $\underline{B} = \underline{e}_z B$ embedded in an infinitely conducting fluid in z < 0 without motions, and in an infinitely conducting fluid in z > h with velocity $\underline{v} = \underline{e}_x v(y)$. In the layer between (0 < z < h) the fluid is a perfect insulator (air). The configuration is sketched in Figure 5. The electric

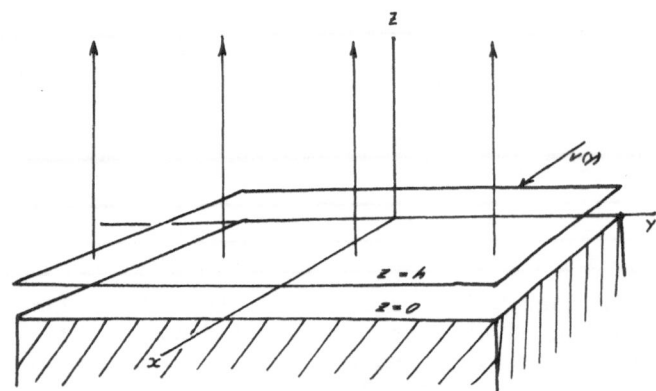

Fig. 5. A sketch of the convecting vertical magnetic field above
the base of the ionosphere (z = h) with the non-conducting atmosphere
(0 < z < h) and the solid conducting earth (z < 0) below.

field in the fixed conductor (z < 0) is zero. The electric field
in the moving conductor (z > h) is

$$\underline{E} = - \underline{v} \times \underline{B}/c$$

$$= \underline{e}_y v(y) B/c \equiv \underline{e}_y E(y).$$

In the gap between (0 < z < h) we have $E = -\nabla \Phi$ and $\nabla^2 \Phi = 0$. The
boundary conditions are $\Phi = 0$ on z = 0 and

$$\Phi(y,h) = \Phi(0,h) - \int_0^y dy' \, E(y')$$

on z = h.

As a first example suppose that v(y), and hence E(y), varies
only very slowly with y, with a characteristic scale ℓ large compared to h. Then Φ has the same characteristic scale and in
0 < z < h,

$$E_y = - \frac{\partial \Phi}{\partial y} = 0 \left[\frac{\Phi - \Phi(0,h)}{\ell} \right] , \quad E_z = - \frac{\partial \Phi}{\partial z} = 0 \left(\frac{\Phi}{h} \right).$$

So $E_y \ll E_z$ and there is a vertical electric field

$$E_z(y,z) \simeq -\Phi(y,h)/h$$

$$= \frac{B}{ch} \int_0^y dy' v(y') - \frac{\Phi(0,h)}{h}$$

across the insulating layer (0 < z < h).

As another example (Walbridge, 1967), suppose that v(y) is
uniform with a value v, and $\Phi(0,h) = 0$. Then $\Phi(y,h) = -vB_y/c$ and

$$\Phi(y,z) = - \frac{vB}{c} \frac{yz}{h}.$$

Hence

$$E_y = \frac{vB}{c} \frac{z}{h} \ , \ E_z = \frac{vB}{c} \frac{y}{h} \ .$$

The vertical component of field grows linearly with horizontal distance y.

In the real situation the boundary z = h between the non-conducting atmosphere and the conducting ionosphere above is indistinct. There is a dense neutral component of gas extending up into the ionosphere (from the atmosphere below) which tends to remain fixed with the atmosphere while the ionosphere convects through it. Collisions produce a strong drag on the ionosphere, causing the ions and electrons to move in a frame of reference in which there are electric fields. High current densities are inferred from the observed magnetic inhomogeneities. These currents flow down along the lines of force into the ionosphere, where they cross over to other lines, and then back out into space. It has been suggested that the combination of high current density and local electric fields cause local particle acceleration.

Now consider the possible role of a plasma sheath or double layer, in a region of intense current (see, for instance, Block, 1972a, b) somewhere at great height (z >> h) where the plasma density is so low that the electron conduction velocities become comparable to the ion thermal velocity. (For instance, a magnetic variation of $\Delta B = 10^{-3}$ gauss over $\ell = 10^2$ km $= 10^7$ cm yields a current density of $j = cB/4\pi\ell = 0.2$ esu. If there are N_e electrons per cm^3, the mean conduction velocity u is $-j/N_e e$, so that for a density of 10^2 electrons per cm^3 (e = 4.8 x 10^{-10} esu) we have u = 5 x 10^6 cm/sec = = 50 km/sec, to be compared with ion thermal velocities of 5 - 10 km/sec (escape velocity is approximately 10 km/sec)). The streaming of the conduction electrons excites plasma turbulence, which introduces a significant impediment to their passage, producing a significant potential drop across the layer. The layer may perhaps develop into a rarefaction wave, becoming a potential double layer in which the current is space charge limited, i.e. the layer is so starved of electrons that a sizeable potential difference must develop merely to accelerate the few electrons available to carry the current. The energy to drive this dissipative condition is, of course, the convective motion and magnetic distortion responsible for the current. It is suggested by the proponents of the idea that other particles (ions and electrons) passing across the layer are then accelerated by the large potential difference that develops. Unfortunately there is no published account, of which I am aware, that sets down all the ideas and shows that the whole scheme is self-consistent. Hence, I must proceed using my own imagination to suppose how the whole thing might work.

First of all, consider the well known elementary theory of the potential double layer - the electrostatic diode - between, say,

$z = a$ and $z = b$. Denote by M_i, $u_i(z)$, and $N_i(z)$ the ion mass, velocity, and density, so that the ion flux is $\phi_i = N_i(z)u_i(z)$ in the upward direction. The electron flux is $\phi_e = N_e(z)u_e(z)$ and the total density is $\rho = (N_i - N_e)e$. The current density is $j = (\phi_i - \phi_e)e$. Under steady conditions conservation of particles requires that ϕ_i and ϕ_e be uniform over z and t. The electric field is in the vertical direction and is related to the potential Φ by

$$E(z) = -d\Phi/dz.$$

The potential Φ satisfies Poisson's equation

$$d^2\Phi/dz^2 = -4\pi e(N_i - N_e).$$

We choose Φ to vanish at the top of the layer ($z = b$) and have a value Φ_0 at the bottom ($z = a$). If the ions and electrons have negligible velocity where they enter (the ions from below, the electrons from above, say), then their velocities are given at a position z within the layer by

$$\frac{1}{2} M_i u_i^2(z) = e\left[\Phi_0 - \Phi(z)\right],$$

$$\frac{1}{2} M_e u_e^2(z) = e\,\Phi(z),$$

and

$$N_i(z) = \phi_i/u_i(z) = \phi_i\left[M_i/2e(\Phi_0 - \Phi)\right]^{1/2},$$

$$N_e(z) = \phi_e/u_e(z) = \phi_e\left[M_e/2e\Phi\right]^{1/2},$$

and

$$\frac{d^2\Phi}{dz^2} = 4\pi e\left[\phi_e\left(\frac{M_e}{2e\Phi}\right)^{1/2} - \phi_i\left(\frac{M_i}{2e(\Phi_0 - \Phi)}\right)^{1/2}\right].$$

Multiply by $d\Phi/dz$ and integrate

$$\left(\frac{d\Phi}{dz}\right)^2 = 8\pi(2e)^{1/2}\left[\phi_e M_e^{1/2}\Phi^{1/2} + \phi_i M_i^{1/2}(\Phi_0 - \Phi)^{1/2}\right].$$

If, as a consequence of the large ion mass, the ion flux is so small that $M_i^{1/2}\phi_i \ll M_e^{1/2}\phi_e$, then

$$d\Phi/dz \simeq (128\pi^2 e\phi_e^2 M_e \Phi)^{1/4}$$

and

$$\Phi = \left[(81\pi^2/2)eM_e\phi_e^2(z-a)^4\right]^{1/3},$$

$$= (81\pi^2 M_e/2e)^{1/3}j^{2/3}(z-a)^{4/3},$$

$$\simeq 10^{-5}j^{2/3}(z-a)^{4/3}.$$

Hence, for the example used above with $j = 0.2$ esu, and $b - a =$ $= 10^5$ cm, say, we have a potential of $\Phi = 10$ stat volts $= 3 \times 10^3$ volts. The electron density in the layer is infinite at the top where the electrons enter (because $u_z = 0$ there) and falls to $0.1/cm^3$ at the bottom where the electron velocity reaches 3×10^9 cm/sec (3 keV). The idea would be, then, that once the double layer is set up, it maintains itself as long as the current density is maintained.

The basic question is how the potential double layer is first set up. Block (1972 a, b) suggests that rarefaction waves may decrease N_e and set up the proper situation. And, how is the layer stabilized?

The most perplexing question for this author is how to fit a possible double layer into the potential field of the convection. The proponents of parallel electric fields and/or double layers as particle acceleration mechanisms in the magnetosphere do not comment on this question in any publication of which I am aware. It is this difficulty that has prevented the growth of my enthusiasm for the idea. Figure 6 is a sketch of my guess as to what might occur. It may well be wrong. It is up to the proponents to do better. The figure shows the plane surface of the earth and the ionosphere a distance h above, with the equipotentials as dashed lines. The magnetic lines of force are vertically upward and not shown (see Figure 5). We imagine that the convection is out of the plane of the page (the x-direction in Figure 5) so that the electric field above the ionosphere is horizontal and to the right. The double layer appears as a horizontal bar across which there is a large

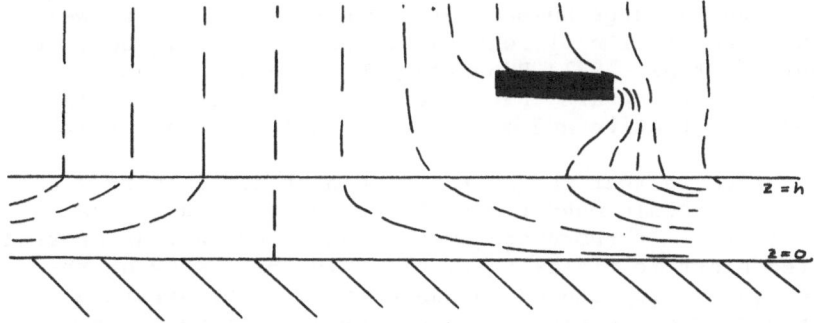

Fig. 6. A guess as to the form of the equipotential surfaces in the neighborhood of a potential double layer (solid bar) in a vertical magnetic field with strong parallel currents.

potential drop. The picture is drawn assuming that there is but little potential drop beneath the double layer (i.e. no convection there). There are other possibilities. The close spacing of the equipotentials to the right of the double layer implies a strong electric field, i.e. very fast convection. Presumably particles are accelerated upward or downward (parallel to B) across the closely spaced equipotential in the double layer.

7. PARTICLE ACCELERATION IN RANDOM ELECTRIC FIELDS

A number of authors have considered the possibilities for particle acceleration in terms of the random electric fields encountered by a particle along its tortuous trajectory through the large-scale magnetic field. Tsytovich (1973) (Kadomtsev and Tsytovich, 1969 and references therein) has been particularly enthusiastic about the general occurrence of plasma turbulence (with frequencies larger than the ion cyclotron frequency) in the field gradients in cosmic magnetic fields and the consequent possibilities for acceleration of particles to high energy. Lerche (1975) has treated the acceleration in electromagnetic waves of random ampli- tude. LeLevier and Marshall-Libby (1968) and Manley and Olbert (1969) have considered particle acceleration in hydromagnetic turbulence (with frequencies below the ion cyclotron frequency). It is observed that the strong collisionless shocks produced by blast waves from solar flares, and by the collision of the solar wind with the magnetosphere of the earth (Parker, 1958, 1965; Parker and Tidman, 1958; Dessler and Fejer, 1963; Bryant, Cline, Desai, and McDonald, 1965; McDonald, Teegarden, Trainor, and von Rosenvinge, 1976), are copious sources of fast particles, presumably accelerated by the strong plasma turbulence generated by the violent impact of the supersonic and subsonic regions of gas. Potential differences of 100 keV applied suddenly across a volume of plasma in the laboratory produce strong currents, instabilities, and turbulence leading to large numbers of 500 keV electrons, as well as some slight acceleration of ions, up to perhaps 5 keV (see, for instance, Wharton et al. 1971; Robertson, Korn, and Wharton, 1972). Altogether it appears that particle acceleration in plasma turbulence can be effective and may play an extensive role in the universe.

The basic idea is that the energy of a high speed charged particle in a large-scale magnetic field is constant insofar as the field is static. If there are small-scale fluctuations present, with associated electric fields $\underline{E}$ ($\underline{r}$, t), then the energy of the particle is perturbed and caused to random walk. The particle experiences a number of accelerations and decelerations as it passes through the random electric field $\underline{E}$ ($\underline{r}$, t). One is as likely as the other if the particle is moving much faster than the phase velocity of the small-scale fluctuations. Thus, if the particle

passes through ν turbulent cells per sec ($\nu = w/\ell$) each with an
instantaneous potential drop of $\pm\delta\Phi$, the effective diffusion coefficient (over energy) is $K = \frac{1}{2}\nu(\delta\Phi)^2$.

The resulting Fokker-Planck equation is readily derived from
elementary considerations on the flux of particles across the energy
T. Denote by $N(T,t)$ the energy distribution of the particles.
Half of the particles at any given energy are projected upward by
an amount $\delta\Phi(T)$ every $1/\nu$ seconds, and half are projected downward.
Thus half the particles in the energy interval $[T - \delta\Phi(T-\delta\Phi), T]$
are projected upward across T every $1/\nu$ seconds. Including terms
second order in $\delta\Phi$ there are $N(T-\frac{1}{2}\delta\Phi(T))\delta\Phi(T-\delta\Phi)$ particles in this
interval, subject to the mean rate $\nu(T-\frac{1}{2}\delta\Phi)$. Hence the net rate
of upward crossings is

$$\frac{1}{2}\left[N(T) - \frac{1}{2}\delta\Phi\partial N/\partial T\right]\left[\delta\Phi(T) - \delta\Phi\partial(\delta\Phi)/\partial T\right]\left[\nu(T) - \frac{1}{2}\delta\Phi\partial\nu/\partial T\right]$$

$$= \frac{1}{2} N\nu\delta\Phi - \frac{1}{4} \partial(N\nu\delta\Phi^2)/\partial T$$

neglecting terms third order in $\delta\Phi$. The rate of downward crossings
from $[T, T + \delta\Phi(T+\delta\Phi)]$ is

$$\frac{1}{2} N\nu\delta\Phi + \frac{1}{4} \partial(N\nu\delta\Phi^2)/\partial T$$

so that the net flux of particles $F(T)$ across T is the difference,

$$F(T) = - \partial(\frac{1}{2} N\nu\delta\Phi^2)/\partial T = - \partial KN/\partial T$$

where $K = \frac{1}{2}\nu\delta\Phi^2$. The negative divergence of $F(T)$ is the rate of
accumulation at T, so that

$$\partial N/\partial t = \partial^2 KN/\partial T^2.$$

We may append any particle losses to this equation. If, for
instance, the particles are subject to a drag force (due either
to collisions or synchrotron radiation) resulting in an energy loss
rate $T/\tau(T)$, the equation becomes

$$\partial N/\partial t = \partial^2 KN/\partial T^2 + \partial(TN/\tau)/\partial T. \tag{27}$$

For stationary conditions (27) reduces to $dF/dT = 0$ or

$$F(T) = - dKN/dT - TN/\tau = \text{constant}$$

which has the general solution

$$K(T)N(T) = \exp\left[- \int_0^T \frac{dT'\, T'}{K(T')\,\tau(T')}\right]$$

$$x \left\{ K(0)N(0) - F \int\limits_{0}^{T} dT' \exp\left[+ \int\limits_{0}^{T''} \frac{dT''' T'''}{K(T''')\tau(T''')} \right] \right\}$$

If all particles pass through the same random fields $\underline{E}$ $(\underline{r},t)$ so that K is independent of energy, and if we suppose further that the characteristic loss time τ is independent of energy, then we obtain the simple result

$$N(T) = \exp\left(- \frac{T^2}{2K\tau} \right) \left[N(0) - \frac{F}{K} \int\limits_{0}^{T} dT' \exp \frac{T'^2}{2K\tau} \right]$$

$$\simeq N(0) \exp\left(- \frac{T^2}{2K\tau} \right) - \frac{F\tau}{T} \left[1 + \frac{K\tau}{T^2} + \ldots + (2n-1)!! \frac{K\tau}{T} + \ldots \right].$$

Ordinarily we expect no source of particles at high energies, so that in the steady state we put $F = 0$. It follows that the energy spectrum is gaussian, with a characteristic width $\Delta T = (2K\tau)^{1/2} = (\nu\tau)^{1/2}\delta\Phi$. The average particle finds itself with an energy $(\nu\tau)^{1/2}$ times the kick $\delta\Phi$ that it gets in each local fluctuation. Suppose, on the other hand, that $\delta\Phi$ is proportional to T, as in the Fermi mechanism. Then, writing $\delta\Phi = qT$ we have $K = \frac{1}{2}\nu q^2 T^2$ so that

$$F = -\tfrac{1}{2}\nu q^2 \frac{\partial}{\partial T} (T^2 F) + \frac{T}{\tau} F.$$

If again the net flux is zero, we obtain a power law

$$F(T) = F(T_0)(T_0/T)^{2(1-1/\nu\tau q^2)}.$$

Equation (27) is presumably equivalent to the Fokker-Planck equation for Fermi acceleration as expressed by Davis (1956) and Parker and Tidman (1958).

More detailed discussion can be found in the references cited above. The work of Tsytovich proposes definite forms for the turbulent spectrum in the plasma, and the consequent particle energy spectra, for special cases such as plasma oscillations (Langmuir waves) and ion acoustic waves. The calculations are, for the most part, limited to the interaction of particles with weak turbulence, employing the quasilinear approximation. The results are interesting but do not establish, so far as we can see, the vigorous acceleration of particles to high energy occurring in nature. Lerche (1975) shows that the acceleration in electro-magnetic waves with phase velocity near c depends on the coherence length of the waves. The larger the coherence length, the more effective the acceleration. It is an observed fact that laboratory plasmas, driven hard by large potential differences supplied from external power sources, produce strong plasma turbulence and a small fraction of the total energy expended goes into very fast particles (Wharton et al. 1971; Robertson, Korn, and Wharton, 1972). But we must be careful in extending

these results to astrophysical circumstances where there is no
external power source. The problem is to produce sufficiently
vigorous plasma turbulence. So far it has not been possible to
demonstrate that the necessary turbulence exists in the cases of
astrophysical interest. Hence we are unable at this time to assess
the effectiveness of plasma turbulence for particle acceleration.
Tsytovich asserts that it is universal and efficient, easily
accounting for the solar flare. Perhaps it is. We have argued
along similar lines (Parker, 1959) for hydromagnetic turbulence.
Certainly an efficient and effective mechanism exists in nature.

8. PARTICLE ACCELERATION BY SUPERNOVAE

The supernova has been a favorite theoretical source of particles
for many years. Supernovae represent an enormous source of energy,
in sufficiently violent form to permit particle acceleration by all
of the mechanisms discussed so far. It is known that supernova
remnants are still producing relativistic electrons, the Crab Nebula
being the most spectracular example visible from earth. A type
II supernova explosion is estimated to involve 10^{51} ergs (Woltjer,
1972). One such explosion in the galaxy every 10^2 years releases
an average of 3×10^{41} ergs/sec, more than sufficient to replenish
the observed cosmic ray energy density in the disk of the galaxy
in the inferred 3×10^6 year cosmic ray life. The supernova is
rich in heavy nuclei, as are the cosmic rays.
       We have already noted, Section 4, that the Fermi mechanism would
seem to function effectively in supernova remnants. Colgate and
Johnson (1970) (Colgate, Grasberger, and White, 1962; Colgate and
White, 1966) have proposed that the outer layers of the original
star are accelerated bodily to relativistic speeds by the explosion,
the individual nuclei in the fluid becoming the cosmic ray particles.
They demonstrate the acceleration to relativistic velocities with
numerical experiments employing a particular model for the opacity
and energy transpost within the matter. The electromagnetic pulse
emitted by the explosion (Colgate, 1975) is an associated phenomenon
that has yet to be fully developed. The idea is intriguing, but
it remains to be shown that matter accelerated from the surface in
this way will in fact contain the heavy elements (believed to exist
in the stellar interior) that make the supernova origin attractive
in the first place. And there is, then, the question of trapping
the accelerated particles in the galactic field before they blast
their way clear out of the galaxy.
       Another idea for particle acceleration in the supernova is
the spinning neutron star − the pulsar (Hewish et al. 1968) −
that remains after the explosion. The rapidly spinning neutron
star with an enormous magnetic moment perpendicular to the spin
axis (Pacini, 1968) emits an electromagnetic wave with a frequency
equal to the angular frequency of the star. The amplitude of the

wave is sufficiently great as to sweep away most of the charged
particles, pushing them back to where the radiation pressure
falls to the value of the ambient background plasma. Within this
wave zone, which may extend far out from the pulsar (depending upon
conditions), a free charge is accelerated outward and may achieve
very high energy. A proton may reach 10^{15} ev or more (Gunn and
Ostriker, 1969, 1970, 1971; Ostriker and Gunn, 1969).

The mechanism is straight forward. Consider a magnetic dipole
of moment M perpendicular to the z-axis and spinning about the
z-axis with an angular velocity Ω. Then M_x = M cos Ωt, M_y = M sin
Ωt, M_z = 0. Each oscillating component of the dipole radiates.
The electric field of the radiation at position $\underline{r}$ = $\underline{n}r$ relative to
an oscillating dipole M cos Ωt at the origin is

$$\underline{E}\ (\underline{r},t) = -(k^2/r)(\underline{n} \times \underline{M})\ \cos(kr - \Omega t)$$

and the magnetic field is

$$\underline{B}\ (\underline{r},t) = +(k^2/r)\ \underline{M} - \underline{n}(\underline{n} \cdot \underline{M})\ \cos(kr - \Omega t)$$

where Ω = kc and $\underline{n}$ is the unit vector in the direction of the position vector.

Thus at a large distance r along the positive x-axis the
spinning dipole produces the fields

$$E_z = -(Mk^2/r)\ \sin(\Omega t - kr)$$

or

$$B_y = -Ez = +B_o(r_o/r)\ \sin(\Omega t - kr)$$

where

$$B_o r_o \equiv Mk^2.$$

Consider, then, a particle with rest mass m, charge q and
four-velocity (dt/ds, d$\underline{r}$/ds) where s is proper time. The equations
of motion are

$$\frac{d}{ds}\left(m\frac{dt}{ds}\right) = + \frac{q}{c^2}\frac{dz}{ds}\,E_z,$$

$$\frac{d}{ds}\left(m\frac{dx}{ds}\right) = - \frac{q}{c^2}\frac{dz}{ds}\,E_y,$$

$$\frac{d}{ds}\left(m\frac{dy}{ds}\right) = 0,$$

$$\frac{d}{ds}\left(m\frac{dz}{ds}\right) = q\left(\frac{dt}{ds}\,E_z + \frac{1}{c}\frac{dx}{ds}\,B_y\right).$$

Since E_z = $-B_y$, the first and second equations yield

$$\frac{d^2t}{ds^2} = \frac{1}{c}\frac{d^2x}{ds^2}$$

which can be integrated once to give

$$\frac{dt}{ds} = \frac{1}{c}\frac{dx}{ds} + 1$$

where the integration constant has been chosen so that the particle starts from rest, i.e. $dx/ds = 0$ only when $dz/ds = dy/ds = 0$, so that $dt/ds = 1$ at that time. With this initial condition the third equation yields $dy/ds = 0$ for all time. The fourth equation can now be written

$$\frac{d}{ds}\left(m\frac{dz}{ds}\right) = -qB_y.$$

Dividing this into the second equation we obtain

$$\frac{d(dx/ds)}{d(dz/ds)} = \frac{1}{c}\frac{dz}{ds},$$

which can be integrated immediately to give

$$\frac{dx}{ds} = \frac{1}{2c}\left(\frac{dz}{ds}\right)^2.$$

Using this to eliminate dz/ds from the second equation, and noting that in the neighborhood of the x-axis the magnetic field is $B_0(r_0/x)\sin(\Omega t - kx)$, we can write

$$\frac{d(dx/ds)^{3/2}}{d\ln x} = -\frac{3qB_0r_0}{(2c)^{1/2}m}\sin(\Omega t - kx).$$

With the intense fields radiated from a pulsar[†] the charged particle (an electron or proton) is quickly accelerated to extreme relativistic energies in the radial direction so that it keeps pace with the wave and $\Omega t - kx$ is approximately constant, with a value θ, say. It follows, then, that a particle introduced at rest at $x = r_0$, subsequently has a velocity

$$\left(\frac{dx}{ds}\right)^{3/2} = \frac{3qB_0r_0\sin\theta}{m(2c)^{1/2}}\ln\left(\frac{x}{r_0}\right)$$

in the radial direction. The particle energy is

[†]Fields of 10^{12} gauss at the surface of the spinning neutron star of radius 10 km yields $M = \frac{1}{2} \times 10^{30}$ gauss cm³, so that a rotation rate of 30/sec yields $k = 6 \times 10^{-9}$ cm⁻¹ ($\lambda = 10^9$ cm). Hence $B_0r_0 = M k^2 = 2 \times 10^{13}$ gauss cm. The amplitude of the magnetic and electric fields is $2 \times 10^{13}/r$ gauss. At the light cylinder ($r_0 = 1.6 \times 10^8$ cm) this is 10^5 statvolts/cm $= 3 \times 10^7$ volts/cm, so that the potential difference over a distance equal to half a wavelength is 10^{16} volts.

$$T = mc^2\left(\frac{dt}{ds} - 1\right),$$

$$= mc^2\left(\frac{1}{c}\frac{dx}{ds}\right),$$

$$= mc^2\left[\frac{3qB_o r_o \sin\theta}{2^{1/2}mc^2}\ln\frac{x}{r_o}\right]^{2/3}.$$

The particle achieves an energy of the order of $(3qB_o r_o \sin\theta/2^{1/2} mc^2)^{2/3}$ times its rest energy, with most of the acceleration in the first wave period. If r_o is the radius of the light cylinder, then typical waves[†] yield $B_o = 10^5$ gauss, and $3qB_o r_o/2^{1/2}mc^2 = 10^7$ for a proton, or $T = 5 \times 10^{13}$ ev. For an electron it is 5×10^{12} ev.

We can conclude that, if individual protons and electrons can be released into the radiation field of a rotating magnetic dipole moment of the magnitude assumed, then they surely will be accelerated efficiently to relativistic energies. Whether particles are released in nature, and what the contribution is to the observed cosmic rays, is an open question. The observed cosmic rays contain a high proportion of heavy nuclei whose presence is not anticipated in a pulsar magnetosphere.

9. CONCLUSION

There are a number of theoretical ideas for particle acceleration. Unfortunately a quantitative judgment of their effectiveness in any particular astrophysical situation depends critically upon the local small-scale details of the turbulent fields. Such information is not readily forthcoming, even for the nearby magnetosphere of the Earth, or the surface of the sun, so we are left in the unsatisfactory position of noting the production of fast particles and being limited to working backwards to speculate on the circumstances that might have produced those particles by one or another of our favorite mechanisms.

Detailed studies (Datlowe, 1971; Kane, 1974) of the particle and X-ray output of solar flares show the complexity of the flare and the associated particle acceleration, evidently involving at least two stages and mechanisms. It is my impression that there is no unique universal form of particle acceleration in the cosmos. There are a few basic principles, perhaps to be found among those mentioned in this article. But in each violent turbulent circumstance the instantaneous combination of field, turbulence, and current density determine the quantitative characteristics of the turbulence and the particle acceleration. The acceleration mechanisms, whatever they are, are remarkably efficient, evidently of the order of 50 percent in the solar flare. We have yet to show that any of our theoretical ideas can function so effectively.

[†]See footnote in the preceeding page.

Altogether it is clear that the fundamental question of particle acceleration is a very complicated subject. It has come a long way; it has yet a long way to go. Its close connection with turbulence would seem to preclude any final quantitative theoretical treatment.

REFERENCES

Alfven, H. and Carlquist, P., 1967, Solar Phys. $\underline{1}$, 220.
Axford, W.I., 1964, Planetary Space Science $\underline{12}$, 45.
Axford, W.I. and Hines, C.O., 1961, Canad. J. Phys. $\underline{39}$, 1433.
Block, L.P., 1972a, Cosmic Electrodyn. $\underline{3}$, 349.
Block, L.P., 1972b, Earth's Magnetospheric Processes (Dordrecht, Holland: D. Reidel Publishing Co.) ed. by B.M. McCormac, p. 258.
Bryant, D.A., Cline, T.L., Desai, U.D. and McDonald, F.B., 1965, Phys. Rev. Letters $\underline{14}$, 481.
Carlquist, P., 1969, Solar Phys. $\underline{7}$, 377.
Chevalier, R.A., 1975, Astrophys. J. (in press).
Colgate, S.A., 1975, Astrophys. J. $\underline{198}$, 439.
Colgate, S.A., Grasberger, W.H. and White, R.H., 1962, J. Phys. Soc. Japan, Vol. 17, Supplement A-III, International Conf. on Cosmic Rays and Earth Storm, Part III.
Colgate, S.A. and Johnson, M.H., 1960, Phys. Rev. Letters $\underline{5}$, 235.
Colgate, S.A. and White, R.H., 1966, Astrophys. J. $\underline{143}$, 626.
Coppi, B. and Friedland, A.B., 1971, Astrophys. J. $\underline{169}$, 379.
Datlowe, D., 1971, Solar Phys. $\underline{17}$, 436.
Davis, L., 1956, Phys. Rev. $\underline{101}$, 351.
Davis, L. and Chang, D.B., 1962, J. Geophys. Res. $\underline{67}$, 2169.
Dessler, A.J. and Fejer, J.A., 1963, Planet. Space Sci. $\underline{11}$, 505.
Dungey, J.W., 1953, Phil. Mag. $\underline{44}$, 725.
Evans, D.S., 1974, J. Geophys. Res. $\underline{79}$, 2853.
Fermi, E., 1949, Phys. Rev. $\underline{75}$, 1169.
Fermi, E., 1954, Astrophys. J. $\underline{119}$, 1.
Fukao, S. and Tsuda, T., 1973b, J. Plasma Phys. $\underline{9}$, 409.
Gunn, J.E. and Ostriker, J.P., 1969, Phys. Rev. Letters $\underline{22}$, 728.
Gunn, J.E. and Ostriker, J.P., 1970, Astrophys. J. $\underline{160}$, 979.
Gunn, J.E. and Ostriker, J.P., 1971, Ibid. $\underline{165}$, 523.
Hamburger, S.M. and Friedman, M., 1968, Phys. Rev. Letters $\underline{21}$, 674.
Hewish, A., Bell, S.J., Pilkington, J.D., Scott, J.D. and Collins, R.A., 1968, Nature $\underline{217}$, 709.
Hoyle, F., 1949, Some Recent Researches in Solar Physics (Cambridge: Cambridge University Press).
Jokipii, J.R., 1968, Astrophys. J. $\underline{152}$, 799.
Kadomtsev, B.B. and Tsytovich, V.N., 1969, Interstellar Gas Dynamics, IAU Symposium No. 39 (Dordrecht, Holland: D. Reidel Publishing Co.) ed. H.J. Habing, p. 108.
Kalinin, Y.G., Kingsep, A.S., Lin, D.N., Ryutov, V.D. and Skoryupin, V.A., 1970, Soviet Phys. JETP $\underline{31}$, 38.

Kane, S.R., 1974, Coronal Disturbances, IAU Symposium No. 57 (Dordrecht, Holland: D. Reidel Publishing Co.) ed. G. Newkirk, p. 105.
Kellogg, P.J., 1959, Nature 183, 1295.
LeLevier, R. and Marshall-Libby, L., 1968, Science 160, 1447.
Lerche, I., 1967, Astrophys. J. 147, 681.
Lerche, I., 1975, Astrophys. Space Sci. 36, 205.
Low, B.C., 1974, Astrophys. J. 189, 353.
Manley, O.P. and Olbert, S., 1969, Astrophys. J. 157, 223.
McDonald, F.B., Teegarden, B.J., Trainor, J.H. and von Rosenvinge, T.T., 1976, Astrophys. J. (submitted for publication).
McIlwain, C.E., 1975, Proceedings Nobel Symposium–Kiruna, Sweden (London: Plenum Press).
Meyer, P., 1969, Annual Rev. Astron. and Astrophys. 7, 1.
Michel, F.C., 1968, J. Geophys. Res. 73, 4135.
Michel, F.C., 1969, Astrophys. J. 157, 1183.
Morrison, P., Olbert, S. and Rossi, B., 1954, Phys. Rev. 94, 440.
Ostriker, J.P. and Gunn, J.E., 1969, Astrophys. J. 157, 1395.
Pacini, F., 1968, Nature 219, 145.
Papadopoulos, K. and Coffey, T., 1974, J. Geophys. Res. 79, 1558.
Parker, E.N., 1957a, Phys. Rev. 107, 830.
Parker, E.N., 1957b, Ibid. 107, 294.
Parker, E.N., 1958, Ibid. 109, 1328; Astrophys. J. 128, 677.
Parker, E.N., 1960, J. Geophys. Res. 65, 3117.
Parker, E.N., 1961, Ibid. 66, 693.
Parker, E.N., 1965, Phys. Rev. Letters 14, 55.
Parker, E.N., 1968, Stars and Stellar Systems, Vol. VII: Nebulae and Interstellar Matter (Chicago: University of Chicago Press) ed. B.M. Middlehurst and L.H. Aller, Chap. 14.
Parker, E.N., 1969, Space Sci. Rev. 9, 651.
Parker, E.N., 1973, Astrophys. J. 180, 247.
Parker, E.N. and Tidman, D.A., 1958, Phys. Rev. 111, 1206; 112, 1048.
Petschek, H.E., 1964, AAS–NASA Symposium on the Physics of Solar Flares, NASA Special Publication SP-50, p. 425.
Priest, E.R., 1973, Astrophys. J. 181, 227.
Riddiford, L. and Butler, S.T., 1952, Phil. Mag. 43, 447.
Robertson, S., Korn, P. and Wharton, C.B., 1972, Report LPS 110, Cornell University.
Scott, J.S. and Chevalier, R.A., 1975, Astrophys. J. Letters 197, L5.
Shapiro, M.M. and Silberberg, R., 1970, Ann. Rev. Nuclear Sci. 20, 323.
Silberberg, R., 1966, Phys. Rev. 148, 1247.
Silk, J., 1975, Astrophys. J. Letters 198, L77.
Smith, D.F. and Priest, E.R., 1972, Astrophys. J. 176, 487.
Sonnerup, B.U.O., 1970, J. Plasma Phys. 4, 161.
Sonnerup, B.U.O., 1971. J. Geophys. Res. 76, 1971.
Speiser, T.W., 1965, J. Geophys. Res. 70, 1717, 4219.
Speiser, T.W., 1967, Ibid. 72, 3919.
Tsytovich, V.N., 1973, Annual Rev. Astron. and Astrophys. 11, 363.
Vasyliunas, V.M., 1975, Rev. Geophys. and Space Phys. 13, 303.

Walbridge, E., 1967, J. Geophys. Res. 72, 5213.
Wharton, C., Korn, P., Prono, D., Robertson, S., Aver, P. and Dum, C.T., 1971, Proceedings Fourth International Conference Plasma Physics and Controlled Nuclear Fusion Research - Madison, Wisconsin, USA (Vienna: International AEC).
Woltjer, L., 1972, Annual Rev. Astron. and Astrophys. 10, 135.
Yeh, T. and Axford, I., 1970, J. Plasma Phys. 4, 207.
Yeh, T. and Dryer, M., 1973, Astrophys. J. 182, 301.

THE BASIC PHYSICAL PROPERTIES OF A GALACTIC MAGNETIC FIELD

E.N. Parker

Department of Astronomy and Astrophysics
University of Chicago, Chicago, Illinois, U.S.A.

1. THE ORIGIN AND NATURE OF THE GALACTIC MAGNETIC FIELD

It seems worthwhile, in this study of the distant active galaxies,
to consider for a moment what is known of the origin and nature
of the magnetic field of our own. It serves as a starting point
for the development of the basic theoretical properties of magnetic
fields of other galaxies.
    Our observational knowledge of the magnetic field of the galaxy
began with the discovery of the polarization of starlight (Hiltner
1949, 1951, 1956) and its interpretation in terms of the alignment
of dust grains by a large-scale magnetic field (Davis and Greenstein
1951). Fermi (1949, 1954) (Chandrasekhar and Fermi 1953) invoked
a hypothetical galactic magnetic field to account for the confinement
of cosmic rays. The observations of polarization indicate that the
mean field is in the azimuthal direction around the disk of the
galaxy, with local fluctuations over dimensions of a few hundred
pc. The magnitude of the fluctuations ΔB is comparable to the mean
field B (Appenzeller 1969; Jokipii and Lerche 1969; Jokipii, Lerche,
and Schommer 1969). The strength of the field is inferred from the
rotation measure $\int ds \cdot BN_e$ of extragalactic radio sources (Gardner,
Morris, and Whiteoak 1969; Gardner and Whiteoak 1969), and of pulsars
within the disk of the galaxy (Manchester 1972, 1974). Then if one
knows the electron density N_e along the line of sight, he can
evaluate the component of $\underline{B}$ in the line of sight $d\underline{s}$. The accepted
field strength is presently about 3×10^{-6} gauss for the mean azimuthal field in the disk, but we should keep in mind that, if N_e
and B are strongly anti-correlated along the line of sight, this
is a serious <u>under</u> estimate, while if N_e and B are strongly correlated and strongly bunched, we may <u>over</u> estimate the mean field.

G. Setti (ed.), The Physics of Non-Thermal Radio Sources, 169–178. All Rights Reserved.
Copyright © 1976 by D. Reidel Publishing Company, Dordrecht-Holland.

In fact one usually assumes, for want of information to the contrary, that N_e and B are uniform along the line of sight. We will show in a moment that $B = 3 \times 10^{-6}$ gauss seems to be consistent with the other information available.

The pressure and the energy density of the galactic magnetic field is $B_\phi^2/8\pi \simeq 0.4 \times 10^{-12}$ dynes/cm^2 and $[B_\phi^2 + (\Delta B)^2]/8\pi \simeq$ $\simeq 0.8 \times 10^{-12}$ dynes/cm^2. The cosmic rays, which we consider here as a tenuous relativistic gas (Parker 1958, 1969) have a pressure P equal to one third of their energy density U, so that $P \simeq 0.5 \times$ $\times 10^{-12}$ dynes/cm^2. The magnetic field is embedded in the inter-stellar gas, with a mean density of the order of 2 hydrogen atoms per cm^3 ($\rho \simeq 3 \times 10^{-24}$ gm/cm^3) in the form of a flattened disk with a half thickness in our neighborhood of the order of 150 pc. The gas density fluctuates enormously, over scales of $10 - 10^3$ pc, and is in a state of disordered motion, with an rms velocity $\langle v^2 \rangle^{1/2}$ of the order of 10 km/sec, over dimensions λ of $10 - 10^3$ pc. Thus the mean kinetic energy density is of the order of $\frac{1}{2}\rho\langle v^2 \rangle \simeq 1.5 \times$ $\times 10^{-12}$ ergs/cm^3 while the Reynolds stresses are $1/3\rho\langle v^2 \rangle \simeq 1 \times$ $\times 10^{-12}$ dynes/cm^2. It is evident, then, that the interstellar "fluid" is a composite medium made up of three components (gas, field, and cosmic rays) each of about the same strength and each with its own special characteristics (Parker 1958, 1969).

The first question to confront us is the origin of the galactic magnetic field. The earliest idea was simply that it is primordial, based on the calculation that the resistive diffusion time is of the order of 10^{25} years: The original fields are still with us because there is no way to get rid of them; ambipolar diffusion may reduce the decay time to 10^{10} years, but this is still comparable to the estimated age of the galaxy. If, however, we suppose that the disordered motions in the interstellar gas involve a mixing of gas throughout the disk of the galaxy, presumably responsible for the mixing of the heavy elements from their sources in the super-novae, etc., we are directed to the conclusion that the field must be similarly mixed. The diffusion coefficient is $\eta \simeq 0.1 v\lambda \simeq 10^2$ cm^2/sec (Radler 1968; Krause 1969; Parker 1971 a, b; 1973 a, b). The diffusion time t across the characteristic scale height $\Lambda \simeq 150$ pc of the gaseous disk is $4\eta t = \Lambda^2$, giving a loss time $t = 2 \times 10^8$ years. This time is so short compared to the age of the galaxy as to force the conclusion that any initial fields must have dis-appeared long ago. The present fields must be a product of present conditions, generated by contemporary dynamo effects.

We have suggested (Parker 1971 a, b, c) that the non-uniform rotation of the galaxy and the turbulent motion of the interstellar gas combine to produce an effective hydromagnetic dynamo. The non-uniform rotation $\Omega(\widetilde{\omega})$ of the gaseous disk of the galaxy shears all radial fields, producing from them the strong azimuthal field B_ϕ that we observe.

$$\left(\frac{\partial}{\partial t} - \eta \Delta^2\right) \underline{B}_\phi = \underline{e}_\phi \widetilde{\omega} \frac{d\Omega}{d\widetilde{\omega}} B_{\widetilde{\omega}} \tag{1}$$

where $\tilde{\omega}$ represents distance from the z-axis of rotation. The turbulent motions perpendicular to the plane of the disk (i.e. rising and falling cells of gas) expand and contract as they move up and down in the background of the disk. We presume that the Coriolis forces cause the expanding and contracting cells to rotate (Steenbeck, Krause, and Radler 1966) producing cyclonic motions in the same way that they are produced in the atmosphere of the earth. The rising and falling cyclonic cells distort the azimuthal field B_ϕ. A rising cell, for instance, lifts up a portion of the azimuthal lines of force, and rotates the raised portion, so that the projection of the lines on a meridional plane of the galaxy is a loop, as sketched in Fig. 1. A loop of flux represents a short segment of vector potential perpendicular to the meridional plane of the galaxy, so that the effect of the cyclonic motion is to build up an azimuthal vector potential (Parker 1955b, 1970; Braginski 1964a, b; 1965; Steenbeck, Krause, and Radler 1966),

$$\left(\frac{\partial}{\partial t} - \eta\nabla^2\right)\underline{A}_\phi = \Gamma\underline{B}_\phi \tag{2}$$

where the coefficient Γ is approximately equal to the cyclonic component of the turbulent velocity if we assume close packed rising and falling cells. Then since $B_{\tilde{\omega}} = -\partial A_\phi/\partial z$, (1) and (2) form a closed system of equations. The solution in an infinite slab of shearing gas with cyclonic turbulence, simulating the broad disk

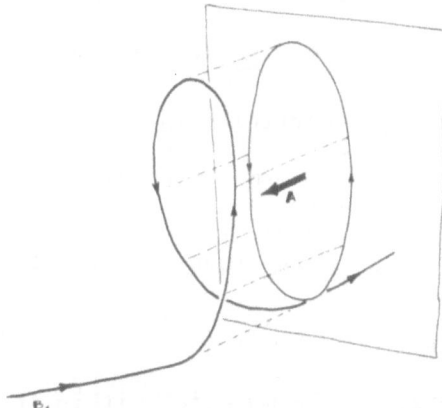

Fig. 1. A sketch of a line of force (of the mean azimuthal field B_ϕ) that is caught in a local rising cyclonic convective cell. The line is carried upward and rotated along with the fluid in the cell, so that its projection on the meridional plane forms a closed loop of field, equivalent to the localized volume of vector potential indicated by the short heavy arrow $\underline{A}$. The subsequent diffusion and superposition of many such loops produces the mean generation of $\underline{A}_\phi$ described by (2).

of the galaxy, indicates strong generation of magnetic field. With
the numbers and signs appropriate to the galaxy, the growth time
of the field is of the order of 2×10^8 years and B_ϕ is an even
function of distance z from the central plane of the disk.   Stix
(1975) has recently carried out a numerical solution, representing
the gaseous disk of the galaxy by a highly flattened oblate spheroid.
Altogether it appears that we may understand the galactic magnetic
field as the child of the non-uniform rotation and cyclonic turbulence of the gaseous disk.   It is an interesting question to what
degree we may extrapolate this conclusion to other galaxies, some
with dense, active gas in their disks, and others with little or
no gas through a nearby spherical volume of stars.   An absence of
gas would seem to imply an absence of magnetic field, as will become clear in the discussion of the next section.

2. THE INSTABILITY OF THE GALACTIC MAGNETIC FIELD

Magnetic fields are expansive, so that their existence in any
astrophysical body requires forcible retention.   The virial equations
for a magnetic field in an isolated system can be written (Chandrasekhar and Fermi 1953; Parker 1954, 1969; Chandrasekhar and Limber
1954; Biermann and Davis 1960),

$$\frac{d^2}{dt^2} \sum m x_i x_j = 2 \sum m \dot{x}_i \dot{x}_j - 2 \int d^3 \underline{r} \, M_{ij} + \sum (x_i F_j + x_j F_i),$$

where the sum is over all the material particles in the body, each
of mass m and subject to a force F_i.   The Maxwell stress tensor is
$M_{ij} = -\delta_{ij} B^2/8\pi + B_i B_j/4\pi$ with the first term representing an isotropic pressure $B^2/8\pi$ and the second term a tension $B^2/4\pi$ along the
lines of force.   For hydrostatic equilibrium of the body, the left
hand side of the equation must be zero.   The trace of the right
hand side then becomes

$$0 = 2T + \int d^3 \underline{r} \, B^2/8\pi + \sum x_i F_i$$

where T is the total kinetic energy of the material within the
body, including the cosmic rays.   It is obvious that the contribution of the magnetic field is positive, being the total magnetic
energy.   The magnetic field contributes a net pressure to the force
balance, just as do the particle motions through their kinetic
energy T.   Hence a magnetic field can be confined only if there are
inward (negative) forces.   Without inward forces the field (as well
as the hot particles) expands and cannot be retained.   There are no
self-confining magnetic fields.   Their internal configuration can
concentrate the expansion in one dimension or another, but the
net effect over all three dimensions is expansion.
     As a consequence of the expansive property of magnetic fields,

it follows that a magnetic field immersed in a gaseous atmosphere is buoyant (Parker 1955a). The field expands until the internal pressure is reduced enough to confine the field, with the result that, with the same temperature inside and outside, the internal density is reduced. The region of field is a bubble that tends to rise. For an isolated flux tube of radius R with a field B along its length, the internal pressure p_i is reduced below the external pressure p_e by the amount

$$p_e - p_i = B^2/8\pi.$$

Hence there is a buoyant force $g(\rho_e - \rho_i)$ per unit volume given by

$$g(\rho_e - \rho_i) = (B^2/8\pi)mg/kT$$
$$= B^2/8\pi\Lambda \tag{3}$$

where T is the temperature, m is the mass per particle, and Λ is the scale height kT/mg of the atmosphere in which the tube is immersed. The flux tube of radius R rises rapidly through the atmosphere (Parker 1974, 1975), with a velocity of the order of $V_A(R/\Lambda)^{1/2}$ where V_A is the Alfven speed.

It is the magnetic bouyancy that is largely responsible for the general instability of magnetic field equilibria. Indeed an equilibrium is unique in itself, existing only when the field is highly symmetric. In the real fields in nature, the field contained within an astrophysical body is not perfectly symmetric, and there is generally no equilibrium at all (Parker 1965, 1969, 1972; Jokipii and Parker 1969; and Yu 1973).

Consider, then, how we are to understand the confinement of the galactic magnetic field to the gaseous disk of the galaxy (Parker 1966, 1968, 1969). The field exerts a strong pressure, seeking to escape. Suppose that the galactic magnetic field is uniform and horizontal (in the azimuthal direction) embedded in a stratified isothermal atmosphere of gas of density ρ and pressure ρu^2 in a gravitational acceleration g, where u^2 is the mean square thermal and/or small scale turbulent velocity in any one direction. Then, including the cosmic ray pressure P, barometric equilibrium requires that

$$\frac{d}{dz}\left(\rho u^2 + \frac{B^2}{8\pi} + P\right) = -\rho(z)g(z). \tag{4}$$

Lacking any detailed information on the variation of field and cosmic rays with height, we suppose that their pressures are proportional to the gas pressure,

$$B^2/8\pi = \alpha\rho u^2, \quad P = \beta\rho u^2$$

where α and β are constants. It follows that

$$\frac{1}{\rho}\frac{d\rho}{dz} = - \frac{g(z)}{u^2(1 + \alpha + \beta)}$$

so that

$$\rho(z) = \rho(o) \exp\left[- \int_{o}^{z} \frac{dz' \, g(z')}{u^2(1 + \alpha + \beta)}\right].$$

Hence one scale height (the height at which the density and pressure has decreased by a factor e) is

$$\Lambda = u^2(1 + \alpha + \beta)/<g>. \tag{5}$$

For u = 6 km/sec, we have $u^2 = 1/3v^2 \simeq 3 \times 10^{11}$ cm^2/sec^2. Then put Λ = 150 pc, and approximate the mean value of g over one scale height as 1.6×10^{-9} cm/sec^2. It follows from (5) that $\alpha + \beta \simeq 1$. This makes sense in view of the currently estimated values of field and cosmic rays (see Section 1). Since

$$P + B^2/8\pi = (\alpha + \beta)\rho u^2 \simeq \rho u^2, \tag{6}$$

and $P + B^2/8\pi \simeq 10^{-12}$ dynes/cm^2, it follows that ρ must equal 3×10^{-24} gm/cm^3, or about two hydrogen atoms per cm^3. Again this is in agreement with current interpretation of the observational evidence.

The point of this exercise is that we must concern ourselves with the confinement of the observed field. The confinement seems to follow in a straight forward way from the gas density and gravitational acceleration inferred from observational data. We must be aware that B, ρ, g, u, and Λ are all intimately related by simple Newtonian mechanics, and we are not free to alter any one of these quantities without having to consider the implications for the others (Parker 1966). A much stronger field, for instance, would demand an increase in ρ and/or Λ. A decrease in ρ would require an increase in Λ or a decrease in B. Thus, within the simple picture of a magnetic field that is largely azimuthal and largely confined to a slab of gas by a gravitational acceleration toward the central plane of the slab, there is an equilibrium for the magnetic field of the galaxy that appears to be consistent with the observed dimensions and densities of the disk.

The question now is whether the equilibrium is stable. It is readily seen (Fig. 2) that it is not, and extensive formal calculation confirms this fact (Parker 1966, 1969). The nature of the instability is evident if we reflect for a moment that the field is buoyant, as pointed out above. The field is held down by the weight of the gas, or, to put it differently, a portion of the weight of the gas is supported by the magnetic field. In picturesque language, the gas rests on the magnetic lines of force. So if the field is perturbed, giving the otherwise horizontal lines of force a slight wavy form, as illustrated in Fig. 2, the gas finds itself resting on a sloping line, with the consequence that it slides downhill

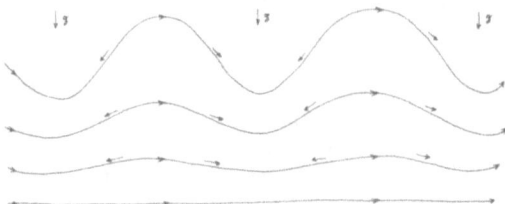

Fig. 2.   A sketch of the perturbed lines of force of a horizontal
magnetic field in a gravitational acceleration g.   The short
arrows indicate the direction of motion of the gas supported by
the field under the force of g.

along the line of force and collects in the low places.   The concentration of gas in the low places is resisted to some degree by
the pressure of the gas, but we expect that the effective γ for the
slow (10^7 years) compression of the interstellar gas is well below
the adiabatic value of 5/3, (Savedoff and Spitzer 1950; Parker 1952,
1953) and the calculations show that any $\gamma < 2$ is insufficient to
block the accumulation of gas (Parker 1966).   The accumulation of
gas weights down the low places and further increases the waviness
of the field.   The cosmic rays, which have pressure, but negligible
weight, enhance the effect by inflating the raised expanded portions
of the field.   The characteristic scale for the instability is a
few times the scale height Λ, say 0.5 kpc.   The characteristic

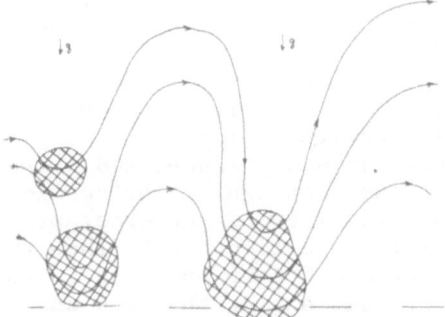

Fig. 3.   A sketch of the general nature of the turbulent inhomogeneous state into which the gas (indicated by cross hatching) is
driven by the instability of the magnetic field, cosmic rays, and
galactic gravitational acceleration.   It must be presumed that the
cosmic rays, continually generated in some way by the stars and gas
at the bottom of the figure, inflate the magnetic arches between
gas condensations, causing the field to expand upward at velocities
of the general order of 50-100 km/sec.

growth time is equal to the scale divided by the Alfven speed, some
0.5×10^8 years. In view of the general instability, we must conclude that the present state of the interstellar medium is a turbulent dynamical state with large variations in gas density as a
consequence of the general instability, something like Fig. 3. It
would appear (Parker 1966, 1969; Mouschovias 1974; Shu 1974; Mouschovias, Shu, and Woodward 1974) that this effect is the principal
architect of the concentration of interstellar gas into large cloud
complexes. Appenzeller (1971) has shown the striking resemblance
of the magnetic field configuration in the α-Persei Cluster to the
theoretical picture, Fig. 3. The clumping of the gas is largely
due, then, to the buoyancy of the magnetic field in the gravitational
acceleration g of the galactic disk. It is more effective, by the
factor g^2/GB^2 than self-gravitation and the associated Jeans instability.

We should remark that the continual generation of cosmic rays
within the disk leads to continual inflation of the magnetic arches
between gas condensations. The 3×10^6 year life τ of the individual
cosmic ray particles in the disk implies a bulk outflow of cosmic
rays at a rate of the order of $\Lambda/\tau \simeq 50$ km/sec through the surface
of the gaseous disk. This, then, is the rate of inflation of the
magnetic arches, or bubbles (Parker 1965, 1966, 1969; Jokipii and
Parker 1969). We have no idea how far the magnetic bubbles may
be extended by the cosmic rays before rapid reconnection, or some
dynamical instability, cuts them free. Fig. 3 is not implausible,
and involves vertical dimensions L comparable to the horizontal
scale, of the general order of 0.1 - 1 kpc. On this basis we expect
that our galaxy must have at least a small halo, or magnetic fuzz,
extending outward from the surface of the gaseous disk. The precise dimensions will have to be determined from observation somehow.
Such a halo may increase the time that cosmic rays are trapped in
the magnetic field of the galaxy from the time $\tau = 3 \times 10^6$ years
spent in the gaseous disk to a time to $\tau L/\Lambda$, of the order of 10^7
years, spent in the halo where the gas density is very small. This
may account for the scarcity of radioactive nuclei Be[10] among the
cosmic rays reported recently by Garcia-Munoz, Mason, and Simpson
(1975) from which they suggest that the mean age of the cosmic ray
particles passing through the solar system is of the order of
2×10^7 years. If this estimate of the age is correct, then the
individual cosmic ray particles circulate freely back and forth
along the magnetic lines of force, in and out of the disk and
through a halo of $L \simeq 1$ kpc.

REFERENCES

Appenzeller, I., 1969, Astrophys. J. 151, 907.
Appenzeller, I., 1971, Astron.&Astrophys. 12, 313.
Biermann, L. and Davis, L., 1960, Zeit Astrophys. 51, 19.
Braginskii, S.I., 1964a, J. Exper. and Theoret. Phys. 47 1084, 2175.

Braginskii, S.I., 1964b, Geomag. i. aeron. 4, 732.
Braginskii, S.I., 1965, Soviet Phys. - JETP 20, 726, 1462.
Chandrasekhar, S. and Fermi, E., 1953, Astrophys. J. 118, 113.
Chandrasekhar, S. and Limber, D.N., 1954, Astrophys. J. 119, 10.
Davis, L. and Greenstein, J., 1951, Astrophys. J. 114, 206.
Fermi, E., 1949, Phys. Rev. 75, 1169.
Fermi, E., 1954, Astrophys. J. 119, 1.
Garcia-Munoz, M., Mason, G.M. and Simpson, J.A., 1975, Paper OG7-7,
14th International Cosmic Ray Conference, Munich.
Gardner, F.F., Morris, D. and Whiteoak, J.B., 1969, Australian J.
Phys. 22, 79, 813.
Gardner, F.F. and Whiteoak, J.B., 1969, Australian J. Phys. 22, 107.
Hiltner, W.A., 1949, Astrophys. J. 109, 471.
Hiltner, W.A., 1951, Ibid. 114, 241.
Hiltner, W.A., 1956, Astrophys. J. Suppl. 2, 389.
Jokipii, J.R. and Lerche, I., 1969, Astrophys. J. 147, 1137.
Jokipii, J.R., Lerche, I. and Schommer, R.A., 1969, Astrophys. J.
Letters 157, L119.
Jokipii, J.R. and Parker, E.N., 1969, Astrophys. J. 155, 777, 799.
Krause, F., 1969, Acta Univ. Wrat. no. 77, p. 157.
Manchester, R.N., 1972, Astrophys. J. 172, 43.
Manchester, R.N., 1974, Ibid. 188, 637.
Mouschovias, T.Ch., 1974, Astrophys. J. 192, 37.
Mouschovias, T.Ch., Shu, F.H. and Woodward, P.R., 1974, Astron.&
Astrophys. 33, 73.
Parker, E.N., 1952, Nature 170, 1030.
Parker, E.N., 1953 Astrophys. J. 117, 169.
Parker, E.N., 1954, Phys. Rev. 96, 1686.
Parker, E.N., 1955a, Astrophys. J. 121, 491.
Parker, E.N., 1955b, Ibid. 122, 293.
Parker, E.N., 1958, Phys. Rev. 109, 1328.
Parker, E.N., 1965, Astrophys. J. 142, 584.
Parker, E.N., 1966, Ibid. 145, 811.
Parker, E.N., 1968, Stars and Stellar Systems Vol. VII, Nebulae
and Interstellar Matter (Chicago: University of Chicago Press) ed.
B.M. Middlehurst and L.H. Aller, Chap. 14.
Parker, E.N., 1969, Space Sci. Rev. 9, 651.
Parker, E.N., 1970, Astrophys. J. 160, 383, 162, 665.
Parker, E.N., 1971a, Ibid. 163, 255, 279.
Parker, E.N., 1971b, Ibid. 166, 295.
Parker, E.N., 1971c, Ibid. 168, 231, 239.
Parker, E.N., 1972, Ibid. 174, 499.
Parker, E.N., 1973a, Astrophys. Space Sci. 22, 279.
Parker, E.N., 1973b, Ibid. 24, 279.
Parker, E.N., 1974, Ibid. 31, 261.
Parker, E.N., 1975, Astrophys. J. 198, 205.
Radler, K.H., 1968, Zeit. f. Naturforsch 23a, 1841.
Savedoff, M. and Spitzer, L., 1950, Astrophys. J. 111, 593.
Shu, F.H., 1974, Astron.&Astrophys. 33, 55.
Steenbeck, W., Krause, F. and Radler, K.H., 1966, Zeit. f.

Naturforsch 21a, 369.
Stix, M., 1975, Astron.&Astrophys. (in press).
Yu, B., 1973, Astrophys. J. 181, 1003.

PULSARS

F. Pacini

ESO, c/o CERN, 1211 Genève 23, Switzerland
and Istituto di Fisica, Università di Roma

1. INTRODUCTION

The discovery of pulsars was announced several years ago (Hewish et al. 1968). One year later the astronomical community became convinced that these objects had to do with the rotation of magnetized neutron stars (Gold 1968; Pacini 1967, 1968). In the meantime the study of pulsars has become an important part of astronomical research. Its importance lies not only in the startling character of the emission but - much more - in the realization that these objects could be the long sought machines which produce in the cosmos relativistic particles and magnetic fields. An understanding of pulsars could be of great importance if one wants to explain other high energy phenomena such as the origin of cosmic rays, the non-thermal activity in galactic nuclei and the nature of galactic x-ray sources.

2. OBSERVATIONAL FACTS

Pulsars are characterized by the emission of flashes of radiowaves at almost exactly maintained time intervals. The mean time between successive pulses is the period P and it is extremely regular. The ratio between pulse width τ and period P defines the duty cycle $\delta = \tau/P$: typically $1 < \delta < 10\%$ with an average value of about 3%. The regular timing

G. Setti (ed.), The Physics of Non-Thermal Radio Sources, 179–196. All Rights Reserved.
Copyright © 1976 by D. Reidel Publishing Company, Dordrecht-Holland.

contrasts with a very irregular behaviour of the intensity which shows changes from pulse to pulse as
well as on time scales up to weeks and months.
    In the following we review the main properties
of pulsars.

1. Period's distribution and timing. The periods
range from 33 msec. for the Crab Nebula pulsar NP 0532
up to 3.75 sec for NP 0527. About 150 sources are
presently known. For more than half of them an accurate timing is available and has led to a measurement
of secular changes. The data have been recently
collected and discussed (Lyne et al. 1975). With no
exception, pulsar periods tend to increase if monitored over a sufficiently long time span. The quantity
$P/\dot{P}$ is then regarded as an age indicator and can range
from about 10^3 years (NP 0531) up to some billion
years (PSR 1952 + 29).
    It is well established that fast pulsars slow
down faster than slow sources. This can account for
the paucity of short period sources in the period's
distribution. On the other hand, the lack of sources
with long periods is probably due to the fact that
their emission becomes too weak for being observable.
    Of particular interest for pulsar electrodynamics
is the question of the slowing-down law. Theoretical
models based upon the idea that neutron stars have a
dipolar magnetic field predict a rotational energy
loss

$$\frac{d}{dt}\left(\tfrac{1}{2}I\Omega^2\right)$$

proportional to the square of the dipole moment M and
to the fourth power of the rotation frequency

$$\Omega = \frac{2\pi}{P}$$

(Pacini 1967, 1968; Goldreich and Julian 1969;
Ostriker and Gunn 1969). In this case

$$\dot{\Omega} \propto \frac{M^2}{I}\,\Omega^3$$

More generally, one can write for the braking law

$$\dot{\Omega} \propto \Omega^n$$

where n is the so-called braking index.

A direct measurement of n has been possible only in
the case of the Crab Nebula and has given n=2.4±0.2
(see e.g. Boynton et al. 1972). The discrepancy with
the simply theory which predicts n = 3 is not serious
and could result from a distortion of the field
geometry in a radial direction (Goldreich et al. 1971).

If all pulsars had the same M^2/I ratio, one would
expect $P \propto P^{-1}$. There is no evidence for such a correlation (Lyne et al. 1975). This is not surprising
because neutron stars, like normal stars, may have
widely different properties at birth. Alternatively,
Lyne et al. (1975) suggest that this may be due to
ohmic decay of the original field: if so the ratio
$P/\dot{P}$ would not indicate any longer the age of a pulsar
as soon as the field has decayed appreciably.

Superimposed on the secular slowing down there
are several instances of sudden period jumps ("glitches").
In all known cases the periods decreased during a jump.
The most spectacular events have occurred in the third
fastest source PSR 0833-45 which has a period P = 89
msec. The fractional changes are

$$\frac{\Delta P}{P} \sim 2 \times 10^{-6} \qquad \text{(February '69)}$$

$$\frac{\Delta P}{P} \sim 2 \times 10^{-9} \qquad \text{(August '71)}$$

In addition, the rate of frequency change increased by
about 1% during the first (and possibly also during the
second) event. Two glitches have been reported for
the Crab Nebula pulsar, where observations also indicate
the presence of a fine structure in the timing measurements (Boynton et al. 1972). In this source the
fractional changes were smaller:

$$\frac{\Delta P}{P} = (9 \pm 2) \times 10^{-9} \quad \text{in September 1969}$$

$$\frac{\Delta P}{P} \sim 1.7 \times 10^{-9} \qquad \text{in October 1971}$$

Finally, glitches occur also in slow pulsars but at a
reduced rate and magnitude. They have been detected
in 3 out of 6 sources monitored over a few years:
typically

$$\frac{\Delta P}{P} \sim 10^{-10} \qquad \text{(Manchester and Taylor 1974)}$$

Suggested mechanisms for the glitches are quite numerous
but none of them has reached uncontroversial consensus
(for a review see Ruderman 1972).

2. Distances and location in the Galaxy.   The same
pulse appears first at high and then at low radio
frequencies.   This is due to the presence of thermal
electrons in the interstellar medium and to the fact
that in such a medium high frequency waves travel
faster than low frequency waves.   The time delay is
given (with obvious notation) by

$$\Delta t = \frac{2\pi e^2}{mc} \left[\frac{1}{\omega_1^2} - \frac{1}{\omega_1^2} \right] \int n_e dl$$

One defines a dispersion measure DM as follows

$$DM = \int n_e dl$$

Observed DM-values are in the range 3-400 pc cm^{-3}.
It is generally agreed that in the interstellar medium
the density of free electrons is around 3×10^{-2} cm^{-3}
and that there are large fluctuations about this value.
Because of these fluctuations the usual method of obtaining pulsar distances from the amount of dispersion
(assuming a definite value for n_e) can lead to incorrect

results in individual cases but can only be used for
statistical purposes.
        If we assume that we have observed all active
pulsars in our neighborhood (there are about 20 sources
whose dispersion suggests a distance less than 500
parsecs), then the total number of galactic pulsars
would be around 20,000.   If however pulsars emit in a
conic beam (and therefore most of them cannot be observed)
this number may have to be multiplied by a factor of
order ten.   A lifetime around 10^7 years would then
correspond to the birth of one pulsar in the galaxy
every $\sim$ 50 years.   This is close to the estimated rate
of supernovae per galaxy but one should stress that
both the beaming and the lifetime factors are very uncertain.
        Pulsars are concentrated along the galactic plane.
High latitude pulsars tend to have a small dispersion
measure and are probably nearby objects, (say d $\leq$ hundreds of parsecs).   At present there is some marginal
evidence that more pulsars may be found at southern

rather than at northern latitudes.   If so, this could
be due to the fact that the sun itself lies ∿ 20 parsecs above the galactic plane.   For the rest, the
distribution of pulsars is very similar to that of SN
Remnants and OB stars but with a larger scale height
∿ 160 pc (Gunn and Ostriker 1970).   The height distribution may derive from pulsar velocities ∿ hundred
km sec^{-1}.   Some evidence for high velocities may have
been found from scintillation studies (Galt and Lyne,
1972).

3. Pulse shapes and intensity variations.   If examined
individually the pulses show a large amount of fine
structure which varies greatly from one pulse to the
next.   The time scale of these structures can be as
short as 10 μsec., thus it implies that the emitting
region has structures with a size ∿ few Km.   If one
adds together many pulses (say, some hundreds of them),
the integrated shape of each source becomes remarkably
stable and can be regarded as the signature of a given
object.   Individual pulses are narrower than this average profile and in many sources they can be seen
randomly located within the profile.   The profile represents therefore a probability distribution for the
occurrence of individual pulses (Taylor and Huguenin
1971).
     In some sources, however, individual pulses move
regularly from one edge to the other of the mean profile ("drifting subpulses").   If two subpulses are
present simultaneously, their spacing is called P_2; P_3
is then the period with which the pattern repeats
itself (i.e. the time interval between successive
crossings by subpulses of a line of constant phase).
One can also define a "phase drift rate" P_2/P_3.
     In addition to the above some pulsars exhibit
periodic intensity changes: for instance PSR 0834+06
shows that over short intervals alternate pulses are
strong and weak.   Some other sources become unobser-
vable for a few pulse periods ("nulling" phenomenon)
and then return to normal power (Baker 1970).
     The morphology of the pulses allows a tentative
classification of the sources (Taylor and Huguenin
(1971).   Type S sources have a simple pulse shape;
type C sources have a complex shape with strong components; type D sources show drifting subpulses.   It is
interesting to notice that all type S sources have
periods P<1 sec.   Type D pulsars can be divided between

those which drift "earlier" (from the trailing to the
leading edge) and those which drift later.   It has
been claimed that there is a correlation between the
direction of drift and the rate of change in period:
pulsars which drift earlier have smaller period changes
(Ritchings and Lyne 1975).

Finally, pulsars show intensity variations over
widely different time intervals, from seconds up to
months.   Pulse-to-pulse intensity changes are highly
correlated over the whole frequency range and must be
intrinsic to the source.   Superimposed upon this
wideband pulse-to-pulse variations are some changes
with timescales $\sim$ minutes to hours which correlate
over a range 0.1-10 MHz.   These are due to interstellar scintillation caused by irregularities in the
intervening medium.   The irregularities scatter the
waves in various directions so that an observer sees
radiation from many different paths: interference
can produce either strong or very weak pulses.   A
remarkable scintillation effect has recently manifest-
ed itself in the Crab pulsar NP 0531.   At the end of
1973 its radio intensity did decrease by a factor 3
and has remained low for about one year (Rankin et al.
1974).   After that the intensity went up to a normal
level.   During the same period the pulse width did
increase by a factor $\sim$ 10.   There can be little doubt
that both phenomena are related to propagation effects
within the Crab Nebula (Lyne and Thorne 1975).

Other instances of pulse smearing due to propagation effects are well known, for instance the gradual
smearing of the Crab Nebula pulses at low frequencies.
As expected, there appears to be some correlation between their occurrence in pulsars and the distance.

4. Brightness, polarization and spectra.   Pulsars are
detected if their flux density is above 10 to 100 mfu.
If one assumes that the radiation is beamed within one-
tenth of a sphere, typical radio luminosities would be
$10^{27}-10^{29}$ ergs sec^{-1} (for NP 0531 $L_{radio} \sim 6 \times 10^{29}$ ergs
sec^{-1}).
Apart from the beaming, the most startling character
of the pulsar radiation is the very high brightness
temperature of the emitting region, far in excess of
any other astronomical source.   Just as an example,
let us consider PSR 1929+10 whose radio output is
$L_{radio} \sim 9 \times 10^{26}$ ergs sec^{-1}.

Since the emitting region cannot be larger than $c\tau$
(τ is the pulse width), one finds an emissivity
$< 3 \times 10^{10}$ ergs cm^{-2} sec^{-1}. This corresponds to a
brightness temperature around 100 MHz $T_b < 10^{22}$°K.
For the Crab pulsar $T_b > 10^{22}$°K, with occasional
peaks $T_b \sim 10^{33}$°K.
These brightness temperatures can only be achieved
by an extremely coherent radiation process. Incoherent processes are indeed subject to the limitation

$$KT_b < mc^2 \gamma$$

which would entail particle energies in excess of
$10^{18} - 10^{30}$ eV. (It is easily seen that particles of
this energy would never radiate primarily at radio
frequencies). In the case of coherent mechanisms
KT_b cannot exceed the total energy of all particles
radiating in phase. The brightness which can be
achieved is then increased by a factor equal to the
number of charges radiating in phase. Two classes
of coherent processes have been considered and they
involve either bunches of particles ("antenna mechanisms") or negative absorption ("maser mechanisms").
At the present time there seems to be no method to
distinguish between these alternatives. If the radiation is due to bunches, the size of the bunch
along the line of sight has to be less than one
wavelength, say in the range 1 - 100 cm. The emitting
region would contain bunches of different size and the
emitted spectrum should reflect the distribution of
sizes (as is well known, the spectra of incoherent
processes reflect the energy distribution of the particles).
     Most pulsars have a radio frequency spectrum
which rapidly declines at high frequencies. In
several sources low frequency cut-off's have been
observed below 100 MHz. It is difficult to obtain
reliable measurements of the spectral index because
of the large flux variations. This has now been
done for several sources and also for different components of the same pulsar. It turns out that about
half of the sources have a spectral index α between
1 and 2, the rest having either $0 < \alpha < 1$ or $\alpha > 2$. (We
recall that for the Crab Nebula $\alpha \sim 0.25$ and that
most optically thin extragalactic radio sources
have $\alpha \sim 0.7$). Different components of a given pulsar may have different spectra.

The polarization characteristics are different
from pulsar to pulsar but, in most cases, individual
pulses are highly polarized.   Linear polarization is
generally stronger than circular.   In the linear case
the position angle may vary regularly through the
pulse; the circular polarization may change sense about
the middle of the pulse.   Both these results are independent of frequency.   Since the polarization
characteristics vary from pulse to pulse, the averaged
pulses are not as highly polarized as the individual
ones.   In most cases the variations appear to be
random.   However, in the presence of drifting subpulses, each subpulse maintains its own polarization
character.

5. Optical and x-ray emission.   The Crab Nebula
pulsar NP 0531 is the only source which has been
detected at other than radio frequencies.   Indeed,
its emission peaks again in the optical or near infra-
red.   The absorption correction is somewhat uncertain
(by $\pm$ 0.2 magnitudes, see Miller 1973) and therefore
also the location of the maximum is uncertain.   The
optical emission shows no absorbtion or emission lines
and corresponds to an output $\sim$ 1000 times larger than
in the radio.   Because of the much higher frequency
the brightness temperature is only $\sim$ 10^{10} ^{0}K and it is
therefore compatible with an incoherent (but obviously
non-thermal) process.   The deep minimum between the
radio and optical spectrum indicates that indeed two
radiation processes are at work, one (coherent) in the
radio and one (incoherent) in the optical.   On the
other hand, the optical and the radio pulses are simultaneous and therefore the two processes appear to
take place in the same emitting region.   The optical
emission extends smoothly up to higher frequencies.
In the soft x-ray region the pulsar spectrum obeys a
power law with an index α = 0.2 $\pm$ 0.1;   at energies
E>10 KeV the spectrum becomes steeper with
α = 1.08 $\pm$ 0.03 (note that the nebular spectrum above
10 KeV is even steeper α = 1.18 $\pm$ 0.03).   The amount
of power radiated by the pulsars in x-rays and γ-rays
is again about 10^2 - 10^3 times larger than in the optical window but the brightness temperature in x-rays
is only around 10^5 ^{0}K.
    It is interesting to notice that the various
components of the pulsar emission behave differently
in different spectral windows.   In the radio, NP 0531

is characterized by a precursor, a main pulse and an
interpulse (some 13 msec later). At optical and
soft x-ray wavelengths there is no evidence for the
precursor and the main pulse dominates over the interpulse. However, around 1 MeV, the interpulse becomes
the dominant feature but it loses again at still higher
energies.

Until recently the second fastest pulsar was PSR
0833-45 with a period P = 89 msec. The best limits
to its optical emission correspond to a magnitude
$m_v > 24.8$. If so and if one assumes (maybe danger-
ously) that the optical output depends only on period,
then

$$L_{opt} \propto P^{-n} \qquad \text{with } n > 8.4.$$

M. Disney has called my attention to a very important
but unpublished result: the erratic timing of PSR 0833
may have led to a positional error. If so, the limits
on the optical emission may lose cogency. Reports of
detection of x-ray pulsations have appeared but have
never been confirmed. On the other hand, a very
strong output of pulsed γ-rays from PSR 0833-45 has
been reported by the SAS-2 investigators (Thompson et
al. 1975). The γ-ray pulses would have the same
period but a different phase than the radio pulses. If
confirmed, this result would imply that the γ-ray
luminosity of NP 0532 is only $\sim$ 10 times larger than
that of PSR 0833-45 while the same ratio in the optical
could be $>2 \times 10^4$.

6. <u>Pulsars and SN Remnants.</u> Only two pulsars are un-
questionably associated with SN Remnants, NP 0531 in
the Crab Nebula and PSR 0833-45 in Vela. In both
cases there is a positional coincidence as well as a
good agreement between the age of the pulsar (from the
slowing down rate) and the estimated age of the Remnant. Some other possible coincidences have been
suggested between PSR 0611+22 and IC 443; PSR 1154-62
and G 296.8-0.3; PSR 2021+51 and HB 21. PSR 0611+22
is $0°.6$ from the center of IC 443 and its timing indicates an age $\sim 10^5$ years. The estimated age of
IC 443 is $\sim$ 60 000 years, its distance $\sim$ 1.4 Kpc. One
could understand why this pulsar is not in the center
of the nebula if its velocity is $\sim$ 150 km sec^{-1}. The
other associations are on more shaky grounds. It is worth
stressing that one does not expect an association between SN Remnants and "slow" pulsars since their timing

suggests an age far greater than the lifetime of SN Remnants.

Finally it is surprising that among the historical SN Remnants only the Crab Nebula has been found to contain a pulsar. This may be due to:
a) lack of sensitivity in the searches conducted so far
b) directionality of the pulsar emission diagram
c) most supernova do not produce pulsars.

7. Pulsars in binary stars. Only one pulsar has been identified in a binary system. Doppler shifts due to orbital motions with periods less than four years would have been easily detected in the timing of most systems. The paucity of pulsars in binary systems could be due to either of the following: a) most binaries are disrupted when one star explodes; b) the lifetime of pulsars in binary systems is short and very soon the release of rotational energy from the pulsar is replaced by accretion of matter from the companion (this would suffocate the pulsar). Apart from these considerations the discovery of the binary pulsar PSR 1913+16 is of great importance (Hulse and Taylor 1974). Its period is only 59 msec and its apparent age seems to be in excess of 10^7 years. The period of the orbit is $\sim$ 8 hours; the eccentricity 0.6; the projected semi-major axis $a \sin i \sim 1$ solar radius; the mass function

$$\frac{(m_2 \sin i)^3}{(m_1 + m_2)^2} = 0.13 \, M_\odot \, ;$$

the radial velocity v $\sim$ 300 km sec^{-1}.

Various considerations (including the lack of periodic changes in the dispersion measure) suggest that also the companion of the pulsar is likely to be a collapsed star. The binary configuration provides an interesting setting in which to test relativity effects. Some of these effects, such as the advance of the perihelion (a few degrees per year) have been already detected or are under investigation.

Because of its short period this pulsar was originally thought to be a possible source of optical pulses. However its dispersion measure suggests a large distance ($\sim$ 6000 pc) and furthermore a large amount of obscuration appears to be present along the line of sight. As to be expected under these conditions, various attempts to identify an optical object have been unsuccessful.

3. PULSAR MODELS

I have had an opportunity to review the basic theory
of pulsars during the 1974 Nato Advanced Study Institute on the "Origin of Cosmic Rays" (Proceedings edited by Osborne and Wolfendale 1975). In the following
I will repeat the main arguments and problems as they
were given in the above reference, in a somewhat more
compact form.

1. Electrodynamics of a Sphere Rotating around the
   Magnetic Axis.

If one connects the pole to the equator of a magnetized
rotating sphere through a nonrotating circuit, an
electromotive force arises and the circuit is traversed
by a current (Faraday experiment).   In a laboratory experiment, say with a sphere of size $\sim$ 10 cm, field $\sim$
10^4 gauss, spinning frequency $\sim 10^3$ sec^{-1}, the difference
of potential between poles and equator is just a few
volts.   In the case of neutron stars (size $\sim 10^6$ cm,
field $\sim 10^{12}$ gauss, rotating $\sim$ 1000 times a second)
the difference of potential can exceed 10^{16} volts.

Indeed, if the sphere is a good electrical conductor, the internal charges redistribute themselves
under the influence of a magnetic force $\dfrac{q(\underline{\Omega} \times \underline{r}) \times \underline{B}}{c}$.
The material becomes polarized and there is an electric
field

$$\underline{E} = -\frac{(\underline{\Omega} \times \underline{r}) \times \underline{B}}{c}$$

Inside the sphere $\underline{E} \cdot \underline{B} = 0$.   Outside one finds $\underline{E} \cdot \underline{B} \neq 0$.
For a neutron star rotating about once per second, the
parallel electric field E_{11} is $\sim 10^{10}$ volts cm^{-1}and the
electric force on a charge largely exceeds the gravitational force.   The outer parts of the stellar surface
cannot be in equilibrium and the particles are shot out
along the magnetic field lines.   In other words, a
neutron star cannot be surrounded by a vacuum (Goldreich and Julian 1969).

The charges drawn off the surface form a magnetosphere around the star.   When equilibrium is establish-
ed, in this magnetosphere $\underline{E} \cdot \underline{B} = 0$ (at least as long
as one can neglect inertial and gravity effects). The
magnetic field lines are equipotentials.

The pulsar magnetosphere can be divided into two regions.   The first region (corotating magnetosphere) contains the field lines that close before a critical distance $R_c = \frac{c}{\Omega}$.   In this region the particles slide along the rigidly rotating field lines.   The charge density is given by the Poisson equation but the density of particles could be much larger.

The rotation of the space charge $(n_- - n_+)$ is equivalent to a toroidal current $j = (n_- - n_+)e\Omega r$. The magnetic field generated by this current is negligibly close to the star but becomes very important at larger distances : at $r \sim R_c$ it becomes comparable to the field produced by the currents inside the star. This entails the necessity of a self-consistent solution where the space charge given by the field is the appropriate field source.   The basic equations of the problem are well established but they have not been solved in realistic cases.

The corotating magnetosphere cannot extend beyond the critical distance R_c because otherwise the velocity Ωr would exceed the speed of light.   The lines of force which close within the light cylinder on the stellar surface have a polar distance from the rotation axis

$$\delta \gtrsim \delta_o = \left[\frac{\Omega r}{c}\right]^{1/2}, \quad \text{typically } \delta \gtrsim 10^{-2} - 10^{-1}$$

radians.

The lines of force which pass beyond the critical distance R_c define the so-called open magnetosphere: in this region there cannot be pure corotation and the plasma escapes freely under the influence of electromagnetic effects.

The potential difference across the <u>polar cap</u> $0 < \delta < \delta_o$ is

$$\Delta \phi \sim \frac{1}{2} \left(\frac{\Omega R}{c}\right)^2 R B_o$$

This potential difference is available for an electrostatic accelaration up to energies of order

$$E_{max} \sim \frac{1}{2} e \Delta\phi \sim 3 \times 10^{12} \frac{R_6^3}{P^2} B_{12} \text{ eV}$$

(The period P is in seconds, B_{12} is in units of 10^{12} gauss, R_6 is in units of 10^6 cm.).

If we assume that the rotation and magnetic axes are parallel, the poles are at lower potential than infinity: the negative charges flow out primarily from the poles and the positive charges from lower latitudes. These poloidal currents lead to a toroidal component of the magnetic field B_t.   Inside the speed of light cylinder $B_t < B_p$ (B_p is the poloidal field).   At $r >> c/\Omega$, B_t becomes the dominant component.

In the original Goldreich-Julian model, it is assumed that $\underline{E} \cdot \underline{B} = 0$ along the open field lines up to infinity.   The field lines act as conductive wires: the potential drop between the axis of rotation and the first open line is frozen into the escaping plasma and the acceleration of particles takes place very far from the pulsar where the model breaks down.   In real life, the assumption $\underline{E} \cdot \underline{B} = 0$ is unlikely to hold in the open magnetosphere.   As a result the particles could be accelerated relatively close to the star.

2. The Oblique Rotator Model

A neutron star rotating about an axis different from the magnetic axis radiates low frequency waves at the basic rotation frequency (Pacini 1967, 1968; Gunn and Ostriker 1969).   The near-zone is similar to the one discussed for an aligned rotator, with an extra complication of time dependency.   At $r >> c/\Omega$ the electromagnetic field becomes a wave field.   The energy loss is of the order

$$I \Omega \dot{\Omega} = \frac{B_c^2}{8\pi} \ c \ 4\pi \left(\frac{c}{\Omega}\right)^2 \approx B_c^2 \ c \ R_c^2$$

where

$$B_c = \frac{B_o R^3}{R_c^3}$$

(This expression is almost identical to the one which can be obtained in the Goldreich-Julian model). Slowing down measurements determine the strength of the magnetic field in the proximity of the light cylinder.   This ranges between 10^6 gauss for the Crab Nebula pulsar (where $R_c \sim 10^8$ cm $\sim 10^2$ stellar radii) down to about 1 gauss for slow pulsars (where $R_c \sim 10^{10}$ cm).   If the field obeys a dipole law in the near-zone, one obtains

$$I\Omega\dot{\Omega} \sim \frac{B_o^2 R^6 \Omega^4}{c^3}$$

i.e. $\dot{\Omega} \, \alpha \, \Omega^3$.

As we have noted earlier, there is observational support for a largely dipole structure of the magnetic field in the near magnetosphere of NP 0532. The small difference between the dipolar and the observed values could indicate a slight radial distortion of the field lines under the pressure of the out-flowing plasma.

With a dipole geometry, the above mentioned values for the magnetic strength at $r \sim R_c$ correspond to the surface fields around 10^{12} gauss.

Beyond the speed of light cylinder the electro-magnetic energy flux should be constant and the magnetic field (by now predominately toroidal) falls off as $1/r$.

In the oblique rotator model, the large-scale magnetic field produced by a pulsar is the magnetic component of a low frequency wave.

It has been shown by various authors that low frequency electromagnetic waves with $f \equiv eB/mc\Omega \gg 1$ accelerate particles very efficiently. The reason for this high efficiency is simple. Since the gyro-frequency eB/mc is much larger than the wave frequency Ω, the particles move in a strong, nearly static, crossed electric and magnetic field. In a very short time ($\ll \Omega^{-1}$) they reach relativistic velocities along the wave propagation and then they ride the wave at constant phase. In the case of a plane wave, a particle exposed to it acquires a Lorentz factor $\gamma = f^2$; in a spherical wave $\gamma = f^{2/3}$. In the Crab Nebula, at the beginning of the wave zone $f \sim 10^{11}$ and therefore the electrons could acquire an energy $\sim 10^{13}$ eV. It is easy to see that a young pulsar could accelerate particles up to the highest energies found in cosmic rays.

In this theory, the particles produced at any given instant are monoenergetic, but it is unclear whether calculations based upon the scheme of vacuum waves interacting with test particles are relevant to a description of the pulsar environment. The answer depends on the density of the circumstellar plasma. First, one wonders whether these low frequency waves can really propagate away from the pulsar. In principle one expects very little - if any - thermal plasma because the waves would exert enough pressure to expel this material (there is indeed observational evidence that the circum-pulsar region in the Crab Nebula has been cleared of ambient plasma). On the other hand,

the pulsar itself continuously ejects particles along
the open field lines and one should worry at least
about these particles.   The usual propagation condi-
tion $\Omega > \omega_p$ (ω_p is the rest plasma frequency) has to
be modified in order to account for the increased mass
of the relativistic electrons: the new condition is
found to be $\Omega > \omega_p/<\gamma>$.   Whether this condition is
satisfied around the Crab pulsar depends on the assumed
rate of particle outflow.   If the number of escaping
charges matches the charge density given by the Poisson
equation, then the waves are probably able to propagate.
If however, the charge separation is small and the
plasma density largely exceeds the charge density (as
some observational evidence seems to indicate) then
the reality of low frequency waves becomes an open
question.

In conclusion, the simple picture of low frequency,
large amplitude waves accelerating particles deserves
great attention and undoubtedly represents a very
attractive possibility, also in view of its application
to other problems (see the lectures by M. Rees in this
volume).   The plasma physics of this problem is presently under intensive investigation by various groups.

3. Emission mechanisms

The most striking features shown by pulsars are:
    1) the small duty cycle of the emission
    2) the extremely high brightness temperature
    3) the polarization characteristics
    4) the "drifting subpulses" phenomenon in many sources
    5) the optical and X-ray emission from NP 0531 but not
       from other sources.
Generally speaking, the models would like to account for
the pulsed character of the emission which should satisfy
all the following requirements:
    a) there should be a privileged sector in the pulsar
       magnetosphere
    b) the radiation of individual particles in this
       sector should be anisotropic.   This is realized
       if the particles move realistically and the
       cone of emission covers an angle $\sim \gamma^{-1}$ around
       the instantaneous velocity.
    c) all particles seen by a given observer have
       velocities collimated within an angle $\sim \gamma^{-1}$.
It is easy to see that the pulsed character of the
emission would be washed out if anyone of the previous
conditions was not satisfied.   Furthermore, condition
b) implies that the radiating particles should have a
Lorentz factor $\gamma \gtrsim 10^2$.

A first class of models invokes an emission process close to the speed of light cylinder. The simplest and historically first of these models is due to T. Gold (1968) who postulated the ejection of streams of plasma from selected hot spots on the star surface. This plasma is forced to corotate with the star with a velocity $v = \Omega r$ by the strong magnetic field. At $r \sim c/\Omega$ the plasma is relativistic and gives rise to a tangential emission beamed in the direction of motion. The emitted spectrum is given by the usual theory of radiation from particles in curved orbits. The typical emitted frequency is $\omega \sim \Omega \gamma^3$ and therefore $\gamma \gtrsim 10^2 - 10^3$ is required to produce radio emission. If $\gamma \gtrsim 10^5 - 10^6$ the spectrum could reach the optical and X-ray bands. Coherent amplification through the antenna mechanism could operate only at radio frequencies since (size bunch) $\gg \lambda_{optical} \gg \lambda_{\gamma-rays}$. Alternative but somewhat similar processes have been analyzed by Smith (1972) and by Pacini and Rees (1970). In the latter paper the origin of the optical and X-ray emission from NP 0531 is also discussed.

The spectral characteristics of the radiation process remain unchanged if the radiation does not arise from corotation at the speed of light but from the sliding of charged particles along the curved field lines of the open magnetosphere. This could take place either at distances $r \sim c/\Omega$ or close to the star. It has been pointed out that the bundle of the open field lines close to the surface subtends a rather small angle and therefore in this framework one could understand why the emission should be pulsed.

A number of models invoke an emission process close to the stellar surface and consider the pulses as a by-product of the extraction and acceleration of particles. In particular Sturrock with various co-workers (see e.g. Roberts and Sturrock 1973) and Ruderman with Sutherland (1975) have tried to spell out the basic physics of the model. As mentioned earlier, if in the proximity of the star E·B is given by the vacuum value, the particles could reach extremely high energies. In the case of the Crab Nebula $E \sim 10^{16}$ eV and the corresponding curvature radiation would be peaked at $\varepsilon_\gamma \sim 10^{12}$ eV. These γ-ray photons would move at an angle with respect to the magnetic field and may annihilate into e^+e^- pairs if the condition $\varepsilon_\gamma B > 4 \times 10^{18}$ is satisfied (ε_γ in eV, B in gauss). It is assumed that the positrons will be turned around by the electric field but the electrons would escape in bunches along the open field lines. Ruderman and Sutherland (1975) in a recent detailed study claim a satisfactory agreement between the

theoretical expectations of this model and the observed
properties of pulsars.

For more details on various models of pulsar
radiation we refer to a very recent summary by Ginzburg
and Zheleznyakov (1975).

4. Conclusion

Many aspects of pulsar theory are either controversial
or unclear (or both).   On the other hand, a large
amount of information is now available about these objects
and we have realized that they operate on a very simple
basic principle, the gradual conversion of gravitational
energy into electromagnetic fields and fast particles.
Much more work will however be required in order to
understand the details of a situation which is well beyond the possibility of laboratory experiments.

REFERENCES

Baker, D.C., 1970, Nature 228, 42.
Boynton, P.E., Groth, E.J., Hutchinson, D.P., Nanos, G.P.,
Partridge, R.B. and Wilkinson, D.T., 1972, Ap. J. 175,
217.
Galt, J.A. and Lyne, A.G., 1972, Mon. Not. Roy. Astron.
Soc. 158, 281.
Ginzburg, V. and Zheleznyakov, V.V., 1975, Ann. Rev.
Astron. and Astrophys. 13.
Gold, T., 1968, Nature 218, 731.
Goldreich, P. and Julian, W.H., 1970, Ap. J. 160, 979.
Goldreich, P., Pacini, F. and Rees, M.J., 1971, Comments
Astronomy Space Physics 3, 185.
Gunn, J.E. and Ostriker, J.P., 1970, Ap. J. 160,
979.
Hewish, A., Bell, S.J., Pilkington, J.D.H., Scott, P.F.
and Collins, R.A., 1968, Nature 217, 709.
Hulse, R.A. and Taylor, J.H., 1975, Ap. J. Lett. 195,
L51.
Lyne, A.G., Ritchings, R.T. and Smith, F.G., 1975, Mon.
Not. Roy. Astron. Soc. 171, 579.
Lyne, A.G. and Thorne, D.J., 1975, Mon. Not. Roy. Astron.
Soc. 172, 97.
Manchester, R.N. and Taylor, D.J., 1974, Ap. J. Lett.
191, L63.
Miller, J.S., 1973, Ap. J. 180, 183.
Osborne, J.L. and Wolfendale, A.W., 1975, Origin of
Cosmic Rays, Dordrecht: D. Reidel Publishing Co.
Ostriker, J.P. and Gunn, J.E., 1969, Ap. J. 157, 1395.
Pacini, F., 1967, Nature 216, 567.
Pacini, F., 1968, Nature 219, 145.

Pacini, F. and Rees, M.J., 1970, Nature 226, 622.
Rankin, J.M., Payne, R.R. and Campbell, D.B., 1974,
Ap. J. Lett. 193, L71.
Richtings, R.T. and Lyne, A.G., 1975, Nature 257, 293.
Roberts, D.H. and Sturrock, P.A., 1973, Ap. J. 181,
161.
Ruderman, M., 1972, Ann. Rev. Astron. and Astrophys.
10, 427.
Smith, F.G., 1972, Repts. Progr. Phys. 35, 339.
Taylor, J.H. and Huguenin, G.R., 1971, Ap. J. 167, 273.
Thompson, D.J., Fichted, C.E., Knitten, D.A. and Ogelman,
H.B., 1975, Ap. J. Lett. 200, L79.

SUPERNOVA REMNANTS

L. Woltjer
European Southern Observatory
c/o CERN
CH-1211 Geneva 23, Switzerland

1. INTRODUCTION

Supernovae play an essential role for the physical conditions in
the interstellar medium, in the chemical evolution of the galaxy,
in the formation of pulsars and possibly also of X-ray sources
and perhaps also in the production of cosmic rays. Each year
several supernovae are discovered in external galaxies, partly
as a result of systematic searches. Also, in our own galaxy
some supernovae have been seen during historical times (SN 1054:
Crab Nebula, SN 1572: Tycho's nova, SN 1604: Kepler's nova;
other somewhat less certain supernovae appeared in 1006 and in
185, while additional ones may have been suggested). In all
cases the supernovae in our galaxy seem to have left an extended
non-thermal radio source. It is now generally believed that
virtually all such radio sources in our galaxy are actually
supernova remnants.

The shape of the radio remnants is sometimes disk-like, but
more usually ring-like. The non-thermal character of the radiation may be inferred from the radio spectrum (spectral indices
between -0.2 and -0.8) or from the presence of radio polarization. Sometimes also the absence of radio recombination lines
is taken as proof of the non-thermal nature. However, without
quantitative measurements its implication is unclear. In many
sources sensitive measurements have detected these lines, where
previously they were believed to be absent. Probably this indicates that in several remnants there is a mixture of thermal and
non-thermal emission; this is not particularly surprising in
view of the expectation that part of the supernovae explode in

G. Setti (ed.), The Physics of Non-Thermal Radio Sources, 197–201. All Rights Reserved.
Copyright © 1976 by D. Reidel Publishing Company, Dordrecht-Holland.

HII regions.

In several remnants optical studies show expanding nebulosities with a characteristic filamentary structure and with spectra characterized in particular by the strength of the [SII] lines relative to H_α. This appears to be mainly an effect due to the excitation by shock waves. Expansion velocities of the filamentary shells range from values as low as 50 km/sec up to the 5000 km/sec velocities measured in Cas A.

In the case of the Crab Nebula also polarized optical continuum radiation (synchrotron radiation) is seen, but this appears to be exceptional. In fact, in many radio remnants there appear to be no visible nebulosities at all, probably because of unfavourable excitation conditions.

X-ray emission appears to be a rather general feature of supernova remnants with more than half of the brighter radio remnants having been detected as X-ray sources. Except in the case of the Crab Nebula, the X-ray spectra tend to be rather soft.

2. DISTANCES AND DISTRIBUTION

In case the expanding filaments form a spherical shell, the distance may be obtained from a comparison of proper motions and radial velocities in the shell. While the method is simple, it is full of pitfalls. For example, in the case of Tycho's remnant we know that the expansion velocity is of the order of 20,000 km/sec; but the one visible filament does not show any significant motion at all. Part of the difficulty may be that if an excitation front passes through a gas cloud, it is not necessary that the gas itself will obtain a substantial velocity. Other distance determinations depend on measurements of radio absorption lines (H, CH_2O,...) combined with models of the motion of the gas in our galaxy or in the case of remnants in the Magellanic Clouds on the supposedly known distances for these.

If one now plots the surface brightness Σ (distance independent!) against the linear radius R for the sources with acceptable distance determinations, one may fit a relation of the form $\Sigma \propto R^{-n}$ to the data with n between 3 and 4. It remains somewhat uncertain to what extent selection effects are important in this. If the Σ - R relation were generally valid, then distances for all remnants may be obtained from measurements of Σ (which then yield R) and the angular diameter.

Simple models show that a relation of this type is not entirely unexpected. If for example we assume that magnetic flux is conserved and that a relativistic electron spectrum of the form $n(E) = K E^{2\alpha-1}$ is in the expanding volume, we have

$$B \propto R^{-2}, \quad K \propto R^{2\alpha} \quad \text{and} \quad F_\nu \propto R^{4\alpha-2}$$

with B, α and F_ν the magnetic field, the spectral index and the emitted flux at frequency ν. With $\alpha = -0.5$ on the average, we would have $\Sigma \propto F/R^2 \propto R^{-6}$. If on the other hand the electrons mainly radiate in a swept up more constant magnetic field, the exponent has a smaller value. Clearly by a judicious choice of models, it should be rather easy to reproduce the relation suggested by observation.

The space distribution of the supernova remnants resembles that of the interstellar gas - at least in the direction transverse to the galactic plane. Of course, this may be purely a selection effect: if there is not much interstellar gas, then the expansion of the remnant is not much decelerated and it fades rapidly. The ages inferred from expansion velocities range from about 250 years for Cas A to more typical values of 10^4-10^5 years for the majority of remnants.

From the space density and the age distribution a formation rate of a few remnants per century is inferred, but in view of the errors this result is quite uncertain. In external galaxies one supernova per three hundred years appears to be about typical, but in certain giant galaxies of types between Sb and Sc the rate is as high as one per thirty years.

To within the uncertainties the rate of formation of pulsars and of supernovae appear to be the same. However, the uncertainties are such that a difference in rates by a factor of ten cannot be completely excluded. Of the 24 supernova remnants with at 1 GHz a flux in excess of 50 Jy, two or three (Crab, Vela X and probably IC 443) have associated pulsars. Taking into account beaming effects this leaves open, but does not prove, the possibility that all supernovae result in both a pulsar and an expanding remnant.

3. THE CRAB NEBULA

Very detailed information is available on the Crab Nebula (SN 1054). The filamentary shell has a mass of about 1 M$_\odot$ (two thirds of which is helium). It expands at about 1000 km/sec on the average, corresponding to a kinetic energy of 10^{49} ergs. If the motion of the filaments is extrapolated backwards at constant velocity, the explosion date would be estimated at about 1150 A.D. This shows that the motion of the shell is accelerated by the pressure of relativistic particles and magnetic fields. Various models of the time variation of the acceleration may be constructed. These show that if the present magnetic field were stronger

than 10^{-3} gauss or the present energy in the form of protons
more than ten times that in the form of electrons, the acceleration would substantially exceed the observed value.

The synchrotron radiation spectrum of the nebula extends
from radio to X-ray wavelengths. At radio wavelengths the spectral index is about -0.25; it gradually steepens to about -1.0
in the X-ray region. The synchrotron photons in the region
1-100 eV may interact with the relativistic electrons responsible
for the far uv which have γ-values of 10^5-10^6 producing Compton
radiation around 10^{12} eV. From the observed upper limits on the
flux of 10^{12} eV photons, an upper limit on the number of electrons
follows. From the observed flux of synchrotron radiation a lower
limit to the magnetic field needed to produce that radiation may
be derived. This lower limit is about 3×10^{-4} gauss, not very
different from the upper limit of 10^{-3} gauss found before. For
field values in that range, the energies in relativistic electrons and in magnetic fields are about equal. With fields of
10^{-3} gauss the life time of the electrons responsible for the
X-ray emission is of the order of a year only. This is qualitatively consistent with recent observations which show that the
X-ray source is smaller than the optical source by a factor of
two about.

The short life time of the X-ray electrons shows that a continuous acceleration process is needed. In fact, an evaluation
of the energy losses by expansion and by synchrotron radiation
shows this to be the case also for the electrons which radiate
at optical wavelength ($\sim$100 GeV). Amplification of the magnetic
field also must take place or at least must have taken place in
the past well after the supernova outburst. The minimum supply
of electron energy may be estimated at $1-2 \times 10^{38}$ ergs/sec.
Undoubtedly, the source of energy is the pulsar. Current neutron
star models suggest that maybe 5×10^{38} ergs/sec in pulsar rotational energy is being dissipated. This shows that the acceleration process must be quite efficient and also that the energy
going into protons cannot much exceed that going into electrons.
The composition of cosmic rays in the Crab Nebula is therefore
quite different from that observed near the sun.

4. OTHER REMNANTS

While the Crab Nebula may be regarded as an expanding cavity
filled with relativistic particles and magnetic fields and contained by the inertia of the filamentary shell, in most other remnants the interaction with surrounding interstellar matter appears
to be of dominant importance.

When an explosion takes place in the interstellar medium,

four main phases may be distinguished in the subsequent evolution. In phase I the expansion is essentially free. In phase
II, the adiabatic phase, the swept up matter is of importance.
The expansion is quite rapid and as a result the matter, which
passes the shock separating the expanding and the undisturbed
material, is heated to a sufficiently high temperature for
radiative cooling to be unimportant. As more and more interstellar matter is swept up, the expansion velocity becomes
smaller. When it drops to about 200 km/sec, radiative cooling
becomes important and phase III, the isothermal phase, begins.
Of course, the isothermality should not be taken too literally:
in the shock there still is a thin, hot region, while far behind
the shock hot material may still be left from phase II. Finally,
in phase IV, the remnant becomes assimilated in the interstellar
medium. Of course, this sequence of events is highly schematic.
In particular, it neglects the effects of instabilities, of
reflected shocks propagating upstream, and most important of all
of inhomogeneities in the interstellar medium. The latter are
conspicuous in Cas A, where fast moving filaments (5000 km/sec)
are interspersed with filaments that are barely moving. These
latter are probably circumstellar dense regions which are
excited by the passing shock. Nevertheless, the simple models
are qualitatively able to account for the shell-like structure
of most supernova remnants, for the near invisibility at optical
wavelengths of many remnants in phase II, for the optical spectra
in phase III remnants, and for the X-ray emission. A quantitative understanding particularly of the X-rays (in terms of thermal bremsstrahlung models) can only be obtained when the effects
of inhomogeneities are taken into account. On the basis of these
models, the total explosion energies may also be estimated from
the observed parameters. Typical values range between 10^{50} and
10^{51} ergs, depending upon the assumptions which are made. It
would seem that the Crab Nebula supernova may have been an event
of comparatively low energy. Perhaps this is part of the reason
that it is such a spectacular object. If it had expanded with
typical initial supernova velocities ($\sim$10,000 km/sec), the
average pulsar generated magnetic field would now have been much
weaker and the compressed interstellar fields in the shell comparatively strong. As a consequence, the activity in the central
regions would be less conspicuous optically, while the radio
structure would more nearly have the characteristic shell-like
appearance of the older remnants.

REFERENCE

Woltjer, L., 1972, Annual Reviews Astron. and Astrophys. 10, 129.

RADIO STARS

Robert M. Hjellming

National Radio Astronomy Observatory
Charlottesville, Virginia, USA

1. UNIQUE ASPECTS OF THE RADIO STAR PROBLEM

Any discussion of the radio emission from stars should begin by
emphasizing certain unique problems. First of all, one must
clarify a semantic confusion introduced into radio astronomy in
the late 1950's when most new radio sources were described as radio
stars. All of these early "radio stars" were eventually identified
with other galactic and extra-galactic objects. The study of true
radio stars, where the radio emission is produced in the atmosphere
of a star, began only in the 1960's. Most of the work on the
subject has, in fact, been carried out in only the last few years.
Because the real information about radio stars is quite new, it
is not surprising that major aspects of the subject are not at all
understood. For this reason these lectures will be organized
mainly around three questions: what is the available observational
information; what physical processes seem to be involved; and what
working hypotheses look potentially fruitful?
　　　How does the radio star problem differ from that of other
radio sources? This is an important question that can be given
at least four general answers. The first is that because events
critical to the source occur in the environs of stars or star
systems, the physics of the problem will include the effects of
deep gravitational potential wells, with the concomitant strong
density gradients. The second is that various versions of
chromospheres, coronae, and in the case of binary systems, stellar
mass exchange, are likely to become involved in the problem.
Thirdly, the stellar radio source always involves smaller distances
and size scales and much shorter time scales for variation than
most other radio sources. Finally, lurking in the background of

G. Setti (ed.), The Physics of Non-Thermal Radio Sources, 203–228. *All Rights Reserved.*
Copyright © 1976 by D. Reidel Publishing Company, Dordrecht-Holland.

the physics of radio stars is the radio sun and its zoo of
puzzling phenomena that may or may not be related to the more
general radio star problem.

     With this preface we can define more precisely what we will
consider to be radio stars for the purpose of these lectures. We
need this definition to avoid semantic confusion with problems
of "star-like" extra-galactic systems or compact planetary nebulae
or HII region phenomena. For our purposes a radio star will be
defined as a very small (less than an arcsecond) radio source,
coincident to within an arcsecond of the position of "star-like"
object with a stellar optical spectrum, which exhibits radio
variability on time scales of minutes to days and weeks.

     We can divide the known radio stars into seven arbitrary but
convenient categories. Six of these will be discussed in varying
degrees of detail. The seven categories are listed in Table I,
together with a number indicating how many are known as of the summer
of 1975.

Table I

Categories and Numbers of Known Radio Stars

Category	Number
1. UV Ceti-type flare stars	6
2. M Super-giants with transient radio emission	3
3. Variable thermal emitters	4
4. Steady thermal emitters	12
5. "Normal" radio binaries	8
6. Radio emitting X-ray stars	5
7. Pulsars	100's

     The special case of the pulsars will not be discussed in
these lectures as they will be dealt with in detail by Prof.
Pacini.

2.   UV CETI-TYPE FLARE STARS

The first pioneering work on true radio stars was carried out by
Lovell (1969) at Jodrell Bank. During the 1960's Lovell, his
co-workers, and a number of Australian radio astronomers carried
out lengthy searches for radio emission from nearby flare stars of
the UV Ceti type. In doing so they were following an analogy based
upon knowing that optical and radio flares on the Sun were frequently
associated; therefore it was logical to search for radio emission
from other stars known to have strong optical flaring. There are

Table II

The UV Ceti-Type Stars Known to be Radio Stars

Name	Spectral Type	Distance
UV Ceti	both dM5.5e	2.7 pc
V371 Orionis	dM3e	40
EV Lacertae	dM4.5e	5
YZ Canis Minoris	dM4.5e	6.6
AD Leonis	dM4.5e	
Wolf 424AB	both dM5.5e	

about 25-30 dMe stars for which such optical flaring is well known, and UV Ceti is the proto-type. Of these, the six shown in Table II have been observed as radio sources.

The radio stars in Table II are relatively nearby objects, they are all in binary systems, and, as shown in the table, two have both components of the binary exhibiting radio flaring.

The primary observed characteristics of the UV Ceti radio flare stars are the following:
1. Radio flaring predominantly at low frequencies (100's of MHz), with frequent flattening of the spectrum during the course of the flare.

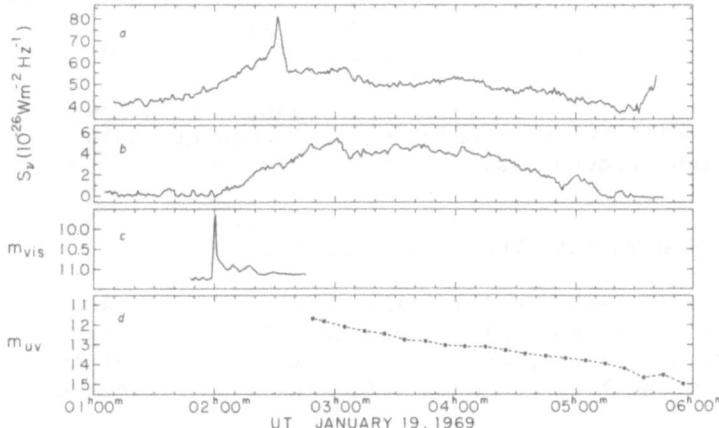

Fig. 1. A radio and optical flare in YZ Canis Minoris as measured at 240 MHz (a) and 408 MHz (b) by Lovell (1969) and at visual (c) and UV (d) wavelengths by Kunkel (1969) and Andrews (1969).

2. Time scales of minutes, sometimes hours.
3. Sharply rising optical "spikes" precede most radio flaring.
4. Frequently a delay between optical and radio peaks, usually a few to several minutes.
5. Drift rates of a few MHz/sec.
6. High brightness temperatures, often up to 10^{15} K.

The interpretation of radio events such as shown in Fig. 1 has been largely based upon modified versions of what happens during large solar radio flares. The most detailed calculations have been carried out by Kahn (1974). In his model one begins by assuming that the star has a chromosphere and corona. The initiating event for the flare is directly associated with optical flaring in the lower chromosphere. This initiating event is assumed to be re-connection of magnetic fields, causing a region of magnetic "heating" which produces a "bubble" of relativistic plasma which rises upwards. After an appropriate time delay, usually a few minutes, the bubble reaches the regions of low enough electron density so that radio emission can escape. The radio emission then proceeds via the synchrotron radiation mechanism, with the bubble rising, the magnetic field strength dropping as the bubble expands, and the radio flux density drifting in frequency and decaying in strength. The parameters of the initiating event are required to release 3×10^{32} ergs of magnetic energy in a surface area 2.5×10^9 cm in diameter, with an initial magnetic field of about 600 gauss. The resulting relativistic particles in the initial bubble must contain 6×10^{30} ergs with a γ of 240. The total radio luminosity of the events is of the order of 3×10^{25} erg/sec.

The high brightness temperatures imply that the synchrotron radiation mechanism must frequently be coherent.

Although we will often be considering the synchrotron radiation mechanism for other radio stars, it is only with the UV Ceti flare stars that the radiation is dominantly at low frequencies. For all the other objects, the dominant radio emission is at high frequencies.

3. M SUPER-GIANTS WITH TRANSIENT RADIO EMISSION

Another type of radio emission initially expected from stars was radio emission from red super-giant coronae. This was predicted by Weymann and Chapman (1965) and a considerable amount of the early searching for stellar radio emission involved nearby red super-giants.

The results to date are tantalizing but extremely inconclusive. Three stars, α Orionis (Betelguese), π Aurigae, and R Aquilae have been reported (Kellermann and Pauliny-Toth 1966, Seaquist 1967, Altenhoff and Wendker 1973, Woodsworth and Hughes 1973) to exhibit very weak, cm-wavelength flaring. Unfortunately, in every case the observations showed positive results for only short periods,

at poor signal to noise ratios, for single antennas operating at
single frequencies. From this we can draw only two useful
conclusions. The first is that, whatever has been observed from
these objects, it has not been the steady coronal emission
originally predicted. The second is that we do not have enough
information to even begin to analyze what may be involved.

4. VARIABLE THERMAL EMITTERS

4.1 The Radio Novae

Three normal novae and one long period variable, R Aquarii, which
has been called a symbiotic nova, have been observed at radio
wavelengths since initial searches began in 1970. The first two
detected were HR Delphini (Nova Del 1967) and FH Serpentis
(Nova Ser 1970) in June 1970 (Hjellming and Wade 1970). These two
novae were observable for the next four years and hence constitute
the largest body of knowledge we have about the radio emission from
these objects. Their radio histories are presented in Fig. 2.
Since the data for R Aqr (Gregory and Seaquist 1974a) and Nova
Scuti 1970 (Herrero et al. 1971) are somewhat fragmentary, the
most useful thing is to focus on the two novae with radio histories
shown in Fig. 2, and in particular on FH Ser for which a nearly
complete radio history is available.
    It is possible to interpret all the data for this class of
objects in terms of thermal bremsstrahlung emission of a hot, ionized

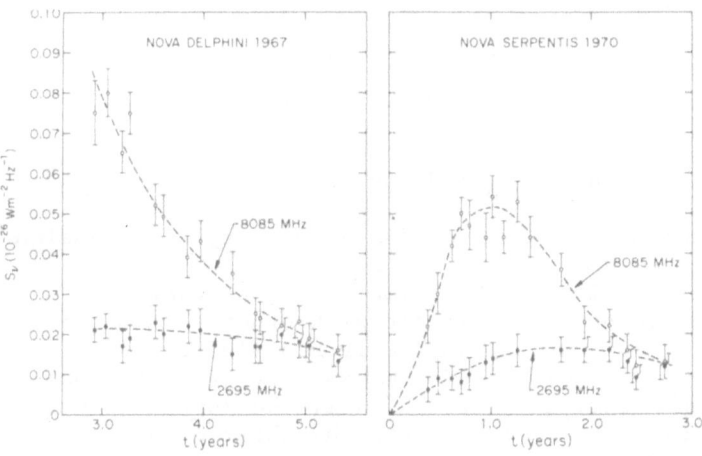

Fig. 2. The radio flux densities at 2695 and 8085 MHz are plotted
as a function of time since initial optical outburst for Nova Del
1967 and Nova Ser 1970.

plasma. To get some idea of the magnitude of the parameters involved, let us discuss a simple uniform source model. If a plasma has a constant electron temperature, T_e, a uniform emission measure, E (the integral of the square of the electron density along the line of sight), over a solid angle Ω_s, it will have a radio flux density, S_ν, at frequency ν, given by

$$S_\nu = (2\ kT_e\ \nu^2)/c^2\ \Omega_s\ (1 - \exp(-\tau_\nu)) \qquad (4.1)$$

where k is the Boltzmann constant, c the speed of light, and

$$\tau_\nu = 0.082\ T_e^{-1.35}\ \nu_{GHz}^{-2.1}\ E \qquad (4.2)$$

where E is in units of pc/cm^6. From examination of eqns. (4.1) and (4.2) we see that given the measurements of the radio flux density at two frequencies, one can solve for two quantities, one being the product of electron temperature and solid angle, and the other being the product of the emission measure and the electron temperature to the -1.35 power. This has the practical limitation with real data that the optical depth must be in the range between 0.1 and 1. Applying this to FH Ser at a time 1.7 years after initial optical outburst, when $S_{2695} = 0.013$ Jy and $S_{8085} = 0.034$ Jy, one obtains

$$T_e\ \Omega_s = 5.8\mathrm{x}10^{-8} \quad \text{and} \quad E\ T_e^{-1.35} = 333 \qquad (4.3)$$

so that, assuming a distance of d=500 pc to the nova, and taking $\Omega_s = \pi\ (D/d)^2/4$, where D is the diameter,

$$T_e\ D_{au}^2 = 7.8 \times 10^8 \qquad . \qquad (4.4)$$

For the $T_e = 10^4$ K which is reasonable, from the optical data, for the nova shell,

$$D_{au} = 280\ \mathrm{au} \quad \text{and} \quad N_e = 2.5 \times 10^5 \qquad . \qquad (4.5)$$

Assuming uniform expansion, this implies a velocity of expansion of 780 km/sec.

The above analysis is somewhat simplified; however, the parameters derived are quite consistent with other information about the nova, and would not be seriously modified by more sophisticated analysis. Beyond this, however, there are a number of points to be kept in mind. First of all, the most interesting result of more detailed analysis is that the observed nova shell must have complex structures. Secondly, as we go on to discuss other radio stars it should be kept in mind that the large, eruptive mass ejection mechanisms of novae may have something in common with mass outflow or mass exchange in other objects. Thirdly, the radio history of the novae involves latter stage evolution from an optically thick thermal source to an optically thin, flat spectrum, thermal source. This seems to be found very frequently in the other radio stars we will discuss. In the case of the novae we cannot rule out a non-thermal mechanism for the observed radio emission and believe the thermal mechanism largely because it is consistent with our conceptions about the nature of the objects.

5. STEADY THERMAL EMITTERS

Something of the order of a dozen stars have compact radio sources that do not appear to change with time and which can be interpreted as thermal sources. These are MWC 349, HBV 475, RY Scuti, V1016 Cygni, HD167362, Vy 2-2, M1-11, MWC 957, P Cygni, NML Cygnus, T Tauri, and Lk Hα-101. All have in common that no radio variability has been seen - as yet. In all cases the thermal emission from gas and dust in surrounding "atmospheres" will explain the radio, infra-red, and optical data. We will not discuss them further in these lectures.

6. RADIO EMITTING X-RAY STARS

6.1 A brief history of radio emission from X-ray stars

The first knowledge that radio emission could be associated with X-ray sources was obtained by Andrew and Purton (1968) and Ables (1969) who found radio emission in the vicinity of the very strong X-ray source called Sco X-1. This object was later found by Hjellming and Wade (1971a) to be a triple radio source with only the central component coincident with the X-ray star.

 The strong variability on time scales of hours found in Sco X-1 suggested that radio counterparts of other X-ray sources could be found by searching the relatively large regions of probable location for these objects as determined from X-ray data alone. In this way Hjellming and Wade (1971b) detected Cyg X-1 and GX17+2 with the NRAO interferometer and Braes and Miley (1971a) independently detected Cyg X-1 with the Westerbork array. Of these two objects Cyg X-1 has achieved a certain notoriety because of the suggestive evidence that the binary system it is associated with may contain a black hole.

 The most spectacular radio star known was found by Braes and Miley (1972) in association with the X-ray source Cyg X-3. Found to attain levels that equal the strongest point sources in the sky, Cyg X-3 will deserve detailed discussion in section 6.4. Finally, the observation of radio emission from Cyg X-2 completes the list of the five compact X-ray sources that can also be described as radio stars. Table III summarizes their properties.

Table III

Properties of Known Radio-emitting X-ray Stars

Name	Star	Radio Characteristics
Sco X-1	faint, blue, 0.787 day period binary, emission lines	non-thermal triple radio source, central component variable on time scales of minutes to hours
Cyg X-1	HDE 226868, close binary, 5.6 day period, O9.7Iab+ black hole ??	weak with flat spectrum most of time, major transition in 1971, transient event in April-May 1975
GX17+2	faint red star	weak, variable, little observed
Cyg X-3	binary IR "star" with 4.8 hr. period	wildly variable, 0.01 to 22 Jy, clear synchrotron source when flarings, time scales hours to day
Cyg X-2	faint, blue, emission lines	weak, "thermal"-like

Most of the information we have about radio emitting X-ray stars has been obtained for Sco X-1, Cyg X-1, and Cyg X-3. For this reason, together with the fact that each of the three behaves in a drastically different manner most of the time, we discuss each of them separately.

6.2 Sco X-1

The radio sources associated with the X-ray star Sco X-1 are arranged in essentially a straight line with the variable component in the middle. A 2695 MHz contour map of these sources made by Wade and Hjellming (1971d) with NRAO interferometer data is shown in Fig. 3.

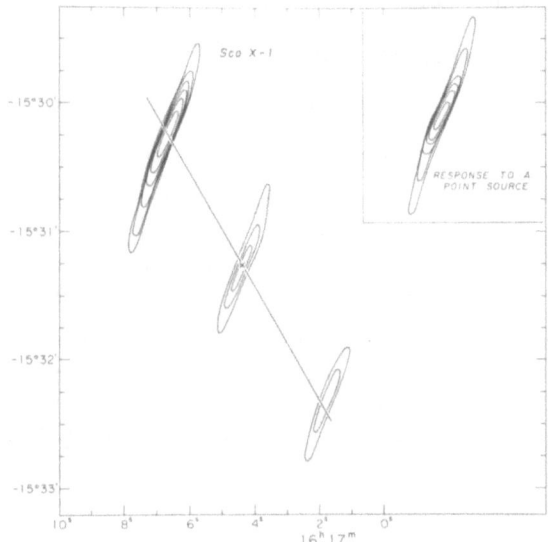

Fig. 3. A map of the Sco X-1 triple radio source at 2695 MHz
together with an indication of the expected response for a point
source. The cross marks the position of the blue star identified
with the X-ray source and the line is drawn to emphasize the
linearity of the configuration.

During the years 1970-1972 the central radio component of Sco X-1
was quite variable and generally dominated the observations.
Thus information about the companion sources was hard to obtain.
However, during the period 1973-1975 the central component, which
previously had been in a dormant state only about 5% of the time,
was found to be in a dormant state roughly 90% of the time.
Table IV summarizes the behavior of the three sources at times
when the central source is not flaring. The 1415 MHz data in Table
IV were obtained by Braes and Miley (1971b) with the Westerbork
array.
      Two major conclusions have been derived from the study of the
weak steady components during 1973-1975. The first is that there
does seem to be a steady radio source on the position of the star
when it is not doing its active flaring. The second is that the
NE and SW companions probably do have structure on size scales of
5 arcsec and 10 arcsec, respectively. This is the reason for the
range of fluxes at 2695 MHz in Table IV. However, definitive
information about this structure must wait until a high resolution
mapping instrument is available that is good at low declinations
like that of Sco X-1.
      The central radio component of Sco X-1 is the only one which
is without question associated with the X-ray star. One could

Table IV

Steady Components of ScoX-1

Component	Location	Frequency	Flux Density
NE	1.2 arcmin from star	1415 MHz	0.019±0.003 Jy
		2695	0.011 to 0.013
		8085	0.004±0.001
Central	on star	2695	0.005±0.001
		8085	0.004±0.001
SW	1.2 arcmin from star	1415	0.017±0.002
		2695	0.004 to 0.008
		8085	< 0.002

argue that the presence of the others is coincidental. Therefore let us now focus on the central radio source.

The star varies in optical wavelengths with a range from roughly 12.4 to 12.8 blue magnitudes. First identified by Sandage et al. (1966), it has recently been found to be a binary system with a period of 0.787 days (Cowley and Crampton 1975). It has cyclic light variations with a range of 0.25 magnitudes, and has been getting fainter since 1965, by roughly 0.3-0.4 magnitudes. Whether this has anything to do with the decrease in frequency of radio flaring is presently simple speculation. The Sco X-1 X-ray source can largely be explained by thermal bremsstrahlung from a plasma with an electron temperature of 4×10^8 K, an electron concentration of 3×10^{16} electrons per cc, and a size of 4×10^7 cm. One of the interesting possibilities for explaining the optical variation is that the X-ray source heats the photosphere of the other star, and the change in aspect with respect to the Earth results in an apparent variation.

The variable radio source associated with the star has always been seen as a variable, non-thermal source. Fig. 4 shows an example of an event observed by Hjellming and Wade (1971d) on Sept. 29-30, 1970.

One of the most interesting characteristics of Sco X-1 events of the type shown in Fig. 4 is the steady flattening of the spectrum during the early rise. Since the interpretation of the radio emission in terms of synchrotron radiation is most probable, this means that the relative proportion of higher energy particles in the radiating regions is increasing, either because the radiating particles are observed while undergoing acceleration, or because

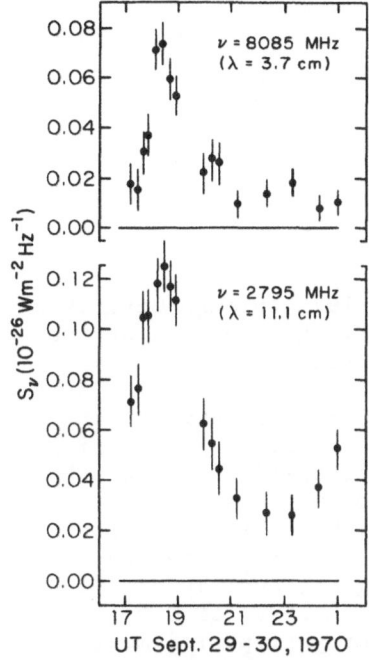

Fig. 4. A radio flare of the central component of the Sco X-1 triple radio source as measured at 2695 and 8085 MHz.

the spectrum of particles injected into the emitting regions is becoming steeper with time.

The closest thing to a model for the Sco X-1 radio source has been developed by Ramaty et al. (1974) who assumed a stellar wind maintained large magnetic field of the order of one gauss out to regions of 3×10^{12} to 3×10^{14} cm from the star. With magnetic fields and emitting regions of this order one can produce the observed quantities of radio emission. This type of model basically shows what parameters are consistent with the radio source data, but does not consider the more interesting questions, such as how are the relativistic particles accelerated in the source, or how one explains the evolution of the events and the frequency of occurence.

6.3 Cyg X-1 (HDE 226868)

The initial discovery of the Cyg X-1 radio source was associated with one of the few events in any X-ray star where a correlation between radio behavior and X-ray behavior exists. Up until March 1971, there was no radio source in the vicinity of Cyg X-1 with upper limits of about 0.005 Jy. At this time the X-ray source was in what has since come to be recognized as one of its higher or

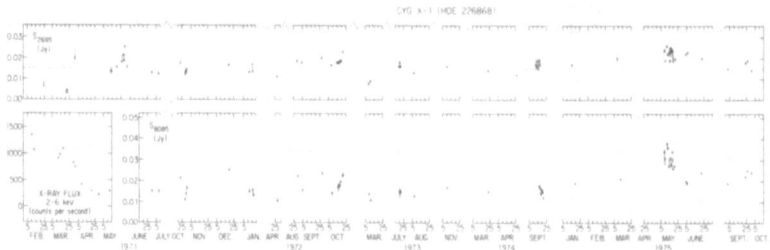

Fig. 5.   The radio history of Cyg X-1 from 1971 through Oct. 1975
as measured at 2695 and 8085 MHz, together with a portion of the
X-ray data for the March 1971 change of state.

stronger states (Tananbaum et al. 1972).   Sometime between
March 22 and 31, 1971, a weak radio source appeared, just during
a time when the X-ray source was making a transition into another
state (Hjellming, 1973).   Some of the radio and X-ray data
showing this is plotted in Fig. 5.   During the four years after
the events of March 1971 the Cyg X-1 X-ray source maintained
roughly the same behavior at X-ray wavelengths, and, as can be
seen from Fig. 5, stayed very close to a mean level of 0.015 Jy
at both radio frequencies.   During this four year period most of
the excitement surrounding Cyg X-1 came as an aftermath of the
identification of the radio source (Hjellming and Wade 1971b,
Wade and Hjellming 1971a, and Braes and Miley 1971a) with the star
HDE 226868.   This system was found to be binary with a 5,599 day
period.   Because it was relatively bright at optical wavelengths,
visual magnitude 9.1, it was eventually shown to be a roughly
20 solar mass O9.7Iab star with a roughly 6 solar mass invisible
companion.   The lack of optical evidence for an object this massive
has led to what can fairly be described as an indictment of the
invisible companion on a charge of being a black hole.
     The flat spectrum radio source associated with Cyg X-1 could
possibly be explained in terms of optically thin thermal
bremsstrahlung emission from a hot plasma.   However, it is not
clear whether we should jump to this simple conclusion.
     As shown in Fig. 5, an unusually high, decaying, radio event
in Cyg X-1 was observed in May-June 1975.   At this time the X-ray
source was also undergoing a transient X-ray event (Holt et al.
1975, Sanford et al. 1975).   During this period there was some
evidence of time variability of the radio source on time scales of
hours (Hjellming et al. 1975) and the source was found to have an
angular size of less than 0.1 arcsec at a time when it was 0.035
Jy at 8085 MHz and 0.022 Jy at 2695 MHz.   If the evidence for time
variation is to believed, it is most unlikely that the radio source
at that time was thermal in origin.   However, if it were, one
can use the same line of argument used for the novae earlier
in these lectures to conclude that a plasma with an electron

temperature of 5×10^5 K, an electron concentration of 3×10^6
electrons per cc, and a size of the order of 250 a.u. could have
been responsible. Interestingly, the excitation parameter derived
based upon this assumption, 13 pc cm^{-2}, is essentially what would
be expected for an ionized region maintained by a O9.7Iab star.

On the other hand, one could easily explain the radio source
in the same qualitative manner discussed for Sco X-1, by
synchrotron radio emission from a region of dimensions 10^{13} cm
with a magnetic field of 10 gauss and an appropriately assumed
supply of relativistic electrons (knowing the distance to be
2500 pc).

It is not clear whether we should think of the Cyg X-1 radio
source in terms of thermal or non-thermal processes, or perhaps
even both at one time or another.

6.4 Cyg X-3

The radio source associated with Cyg X-3 was initially detected
by Braes and Miley (1972) with the Westerbork array. From May
through August 1972 it was an interesting but not spectacular
radio source that flared up to levels as high as 0.5 Jy (Hjellming
et al. 1972a). On August 30-31, 1972 it existed at the very low
levels of 0.01 Jy, only to be accidentally found on September 2,
1972 to reach levels of 22 Jy (Gregory et al. 1972a). Fig. 6
shows observations of Cyg X-3 made by a large number of individuals
at a number of radio observatories during the period Aug.-Oct.,
1972. During and after the events of Sept.-Oct., 1972 Cyg X-3
has had an exciting history as a radio, X-ray, and even an infra-
red source (Becklin et al. 1972). It is invisible at optical
wavelengths, presumeably because of the enormous obscuration known
in the Cygnus region. The X-ray source has been found to have a
4.8 hour periodic modulation (Parsignault et al. 1972) which is
synchronized with a similar 4.8 hour periodicity in the infra-red

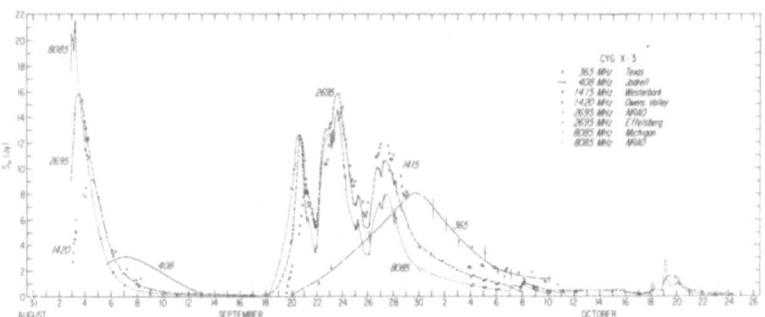

Fig. 6. Multi-frequency radio observations of Cyg X-3 as obtained
by a number of radio observatories plotted as a function of time
during the period Aug.-Oct., 1972.

emission (Becklin et al. 1973), 21-cm absorption experiments
against the variable radio source (Lanque et al. 1972, Chu and
Bieging 1973) have been used to determine a distance of 10,000 pc.

Since 1972 Cyg X-3 has been subjected to extensive theoretical
interpretation and a number of important radio properties have been
established. Most of the time Cyg X-3 has exhibited very frequent
flaring up to levels of a few to ten Jy; however, usually the
observed behavior is so complicated as to be nearly impossible to
interpret. However, an event observed in May 1973 by Seaquist
et al. (1974) was found to have evolving linear polarization of up
to 14% with a surprisingly stable position angle. This strongly
re-enforced the synchrotron radiation interpretation of the events
and established that variable amounts of Faraday rotation due to
plasma associated with the relativistic particles must be responsible
for the varying amount of polarization. The other observational
result of major interest was the discovery that Cyg X-3 can lapse
into extended periods of relatively quiet behavior. An extended
campaign at radio, infra-red, and X-ray wavelengths during Sept.
1974 showed a startling lack of the "normal" Cyg X-3 flaring
(Mason et al. 1976). Fig. 7 shows three weeks of the radio data
gathered on the source during Sept. 1974 using the NRAO interferometer
It was recognized at the time of the Sept. 1972 flaring events
(Gregory et al. 1972b, Hjellming and Balick, 1972) that the data
for Sept. 1-14, 1972 showed obvious signs of the classic behavior

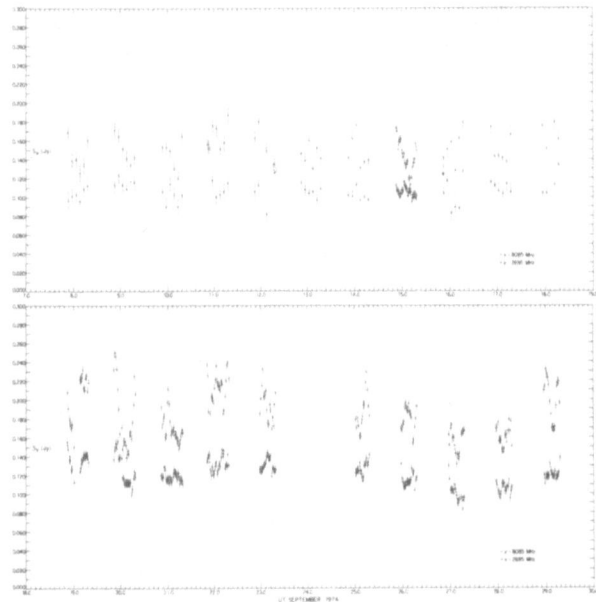

Fig. 7. Radio measurements of Cyg X-3 at 2695 and 8085 MHz plotted
as a function of time for the period Sept. 8-29, 1974.

of an expanding synchrotron source suffering mainly from adiabatic
losses. The latter part of the decay of the Sept. 2-14 event could
be fit very well (Gregory et al. 1972b) to a power law decay
according to the formula

$$S_\nu = 4200 \ (t - t_0)^{-4.1} \ \nu_{GHz}^{-0.55} \ Jy$$

where t is measured in days and t_0 is taken as Sept. 1.3 UT. It
was pointed out by Aller and Dent (1972) that the early part of the
decay of the event could be fit very well by an exponential.
Subsequent work by Peterson (1973), Gregory and Seaquist (1974b),
Hjellming et al. (1974), and Marscher and Brown (1975) have resulted
in a fairly consistent and interesting picture for the large Cyg X-3
flare events. Starting in regions of dimensions of the order of
10^{14} cm, a large number of relativistic particles suddenly find
themselves co-existing with relatively large magnetic fields (tenths
of a gauss to tens of gauss) and varying densities of plasma. The
result is expansion with a combination of adiabatic and synchrotron
losses for the radiating particles. The observed radio source
evolves from a self-absorbed source to an optically thin one with
an interesting intermediate state where the source appears to decay
exponentially because of the effect of synchrotron losses (Marscher
and Brown 1975), only to eventually reach the classic adiabatically
expanding state. Depending on whether the initial thermal plasma
density is low or high, the Faraday rotation within the source is low
or high, and the escaping radiation is un-polarized or linearly
polarized.
      Only a few of the possible scenarios for an evolving synchrotron
radio source of this type have yet been worked out.

7.   "NORMAL" RADIO BINARIES

7.1 Summary and History of the "Normal" Radio Binaries

Before going into the well-studied objects in detail, let us define
the characteristics of this class of radio stars and list their major
known properties. For our purposes we define the class of "normal"
radio binaries as possessing the following four characteristics:
(1) they exhibit variable radio emission mainly at the higher,
GHz-range, frequencies; (2) the time scale for variability is minutes
to hours; (3) they are associated with double or multiple star
systems; and (4) they are not strong X-ray sources.
      As of the summer of 1975 there are eight radio stars belonging
to this class. These eight radio stars and their known properties
to date are summarized in Table V. In Table V one column, L(star)/
L(Sun), is presented as an approximate estimate of the relative radio
luminosities of the strongest events seen on these objects relative
to the strongest events at comparable wavelengths observed on the
Sun. This is mainly to indicate that the energies involved, though

small in themselves, are many orders of magnitude greater than
found in radio events on the Sun.

Table V

Properties of the Known "Normal" Radio Binaries

Name	Spectral Types	Period (days)	Distance (pc)	L(star)/L(Sun)
α Sco B	B4V+M2Ia	>4000	180	3.6×10^4
β Per (Algol)	B8V+G5III +A7m	2.867	25	6.6×10^5
β Lyr	B8p+ ??	12.932	260	1.6×10^5
b Per	A2IV+G +F5V	1.527	56	3.8×10^3
AR Lac	G0IV-V +K0IV	1.983	47	3.0×10^4
CC Cas	O9IV+O9IV	3.369	1000	1.0×10^6
UX Ari	G5V+K0IV	6.438	55	2.5×10^4
RT Lac	G9IV+K1IV	5.074	205	1.2×10^5

Let us now briefly summarize the history of the observations
of the eight objects listed in Table V. In doing so we can
briefly summarize their radio properties, leaving more detailed
discussion of the most important member of this class until later.
   The first member of this class to be discovered was what
eventually turned out to be α Sco B. It was initially detected
as part of a search for radio emission from red supergiants (Wade
and Hjellming 1971b), and the red supergiant Antares, or α Sco A,
was a prime suspect. Initially detected as a weak non-thermal
radio source, the initial observations had insufficient resolution
to tell whether the radio source was associated with the red
supergiant or its companion star 3.2 arcsec away. On June 1, 1971
a flare at 8 GHz was observed and the resulting data showed that the
radio source was associated with the B4V companion, α Sco B
(Hjellming and Wade 1971c). Since that time the importance of this
radio star has been largely due to its influence on subsequent radio
star observing. At that time there was no obvious reason why a
B4V companion to a red supergiant should be a radio source. However,
there were two special things known about α Sco B: (1) it was a
companion to a star known to have regular loss of mass, hence
accretion of some of this matter onto the B4V star was a virtual

certainty; and (2) α Sco B was known to be surrounded by a faint,
extended, irregular nebulosity exhibiting only forbidden emission
lines of FeII and NiII . The obvious thing suggested by these
facts was that perhaps binaries with known mass exchange and
emission lines would be good prospects for radio star emission.
In Oct. 1971, the nearby binaries Capella and Algol were observed
with negative results for Capella and positive results for Algol
(Wade and Hjellming 1972). A few days later β Lyrae was also
detected. Both Algol and β Lyrae were at that time quite weak
(0.01 - 0.02 Jy) and were stronger at 8 GHz than at 2.7 GHz.
Extensive plans were made to hunt for radio emission from many
other Algol- and β Lyr-type binaries and to monitor these objects
themselves. The result of this work in Jan.-Feb., 1972 was a
series of surprises (Hjellming et al. 1972b). First of all, of the
roughly thirty other binaries observed, none were caught as radio
sources. Secondly, β Lyrae had disappeared as a radio source by
dropping below detection limits. Thirdly, the biggest surprise of
all was that Algol exhibited a wide range of variability, ranging
from 0.01 Jy up to 0.32 Jy in just a few hours. The history of
some of these observations is shown in Fig. 8. Aside from the
historical interest of the data in Fig. 8, a number of things
that have been well established in subsequent years are seen.
First of all, there is no obvious correlation of radio flaring
and optical cycle. Secondly, a typical event lasts a fraction of
a day and hence one typically observes only fragments of interesting
events. Thirdly, lacking a predictive theory, chance seems to
determine what is obtained. More details on radio emission from

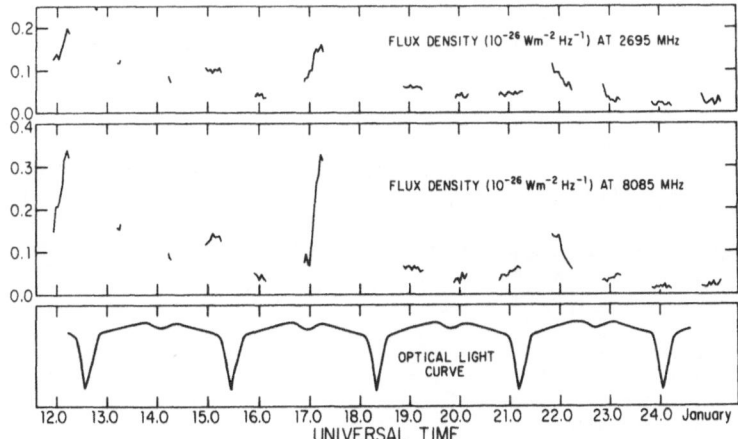

Fig. 8. Radio observations of β Per (Algol) at 2695 and 8085 MHz
made with the NRAO interferometer from Jan. 12 through Jan. 25,
1972, together with a schematic of the corresponding optical light
curve.

Algol will be discussed later since it is by far the best studied "normal" radio binary.

The next observed radio star was b Persei (Hjellming and Wade 1973). This star was put into the search list simply because it was the proto-type of ellipsoidal variables, and at that time many proto-types of classes of stars were being observed. The b Per radio source has always been found to be very weak and somewhat stronger at 8 GHz than at 2.7 GHz.

In Feb. 1973 a search for radio emission from binary stars known to have period jumps or evolving periods (as do Algol and β Lyr, respectively), resulted in the detection of radio emission from AR Lacertae (Hjellming and Blankenship 1973). AR Lac has since become one of the more important radio stars because it has been found to exhibit fairly strong radio events, and also because it was a link in a chain of clues leading to a number of other important radio stars.

The strongest event observed to date in AR Lac (Gibson and Hjellming 1974) is shown in Fig. 9. Fig. 9 shows what is probably a single event lasting roughly one day with a very sharp rise and a slower decay. The radio flux is always strongest at the higher frequency, as is typical of most "flaring" events in the "normal" radio binaries.

The next radio star detected was CC Cas (Gibson and Hjellming 1974). A member of a group of binaries with large mass components, CC Cas has always been observed as a very weak, but variable radio source which is stronger at the higher frequencies. Despite its weakness, its distance makes it intrinsically the most luminous of all the objects listed in Table V.

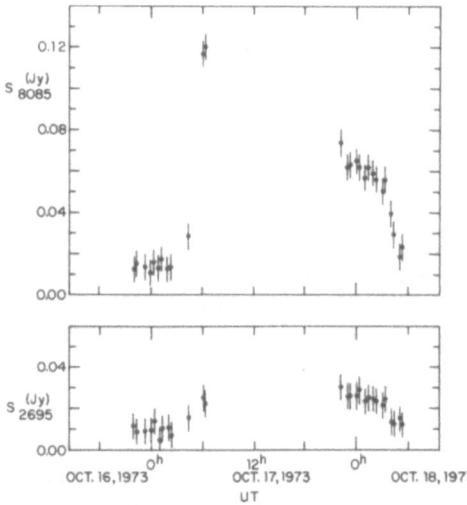

Fig. 9. Radio variations of AR Lac at 2695 and 8085 MHz as observed with the NRAO interferometer on Oct. 16-18, 1973.

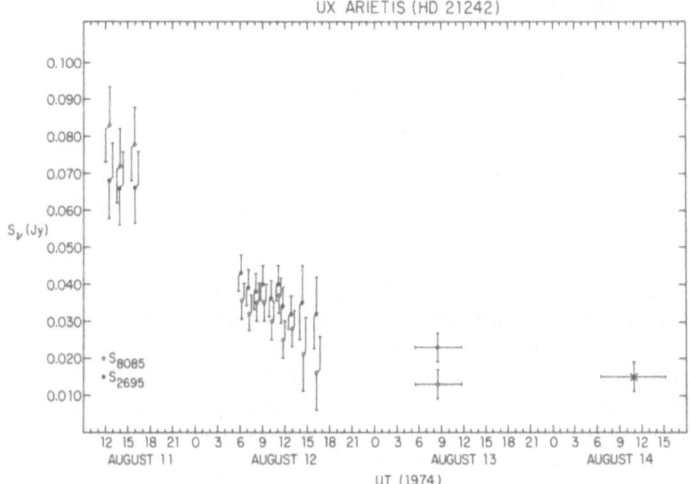

Fig. 10. Radio measurements of an event in UX Ari observed Aug. 11-14, 1974.

Since 1970 a large number of people have suggested their favorite objects were odd enough that they should be radio stars. Many dozens of times they have been wrong, and so far only once has a suggestion from an optical astronomer led to the detection of a radio star. Based upon similarities between AR Lac and UX Ari (HD 21242), D. Hall suggested that UX Ari should be a radio star also. UX Ari was found to be a radio star by Gibson and Hjellming (1975), and an example of one of its stronger flares is shown in Fig. 10. The event shown in Fig. 10 is of special interest because it is one of the very few known in the "normal" radio binaries where a decaying event, initially stronger at the higher frequency, "turns over" and finishes as a decaying non-thermal source.

The most recently detected "normal" radio binary is RT Lac (Gibson et al. 1976). It exhibits Algol-like radio flaring to modest levels and is most important because it is the third of a class of about thirty objects that have been detected as radio stars. The suggestion from D. Hall that UX Ari was a good prospect for radio star emission was eventually broadened into the suggestion that the so-called RS Cvn-class of objects, of which AR Lac and UX Ari are members, should be radio stars (Gibson 1975). RT Lac is a member of this class which also exhibits "Algol-like" radio events. In some ways this may bring us full circle back to some commonality with the red dwarf flare stars and the radio Sun. This is because one of the major characteristics of the RS Cvn-class of objects is the presence of "waves of darkening" and other features that are commonly interpreted as being due to large "star-spots". The importance of these associations is still to be established.

7.2 β Persei (Algol) - The "Proto-type" Radio Binary

Algol as a radio star has become something of the proto-type of
the radio binaries because it has been extensively observed, it has
achieved the highest levels of flaring, and the size scale of the
flaring radio source has actually been measured for two separate
events via VLBI observations. The flaring history of Algol can
be seen in capsule form in Fig. 11. Note the miniscule levels
first found in 1971 and variation in flaring range since that time.
The strongest event, just over 1 Jy, occured Jan. 16, 1975.

Fig. 11 shows that in some sense Algol does have variations
in level of flaring activity. Whether or not these variations
are part of any cycle analogous to the solar cycle will take many
years of observations to prove or disprove. The fact that such
variations in flaring level do exist may help explain the general
result that at any one time most binary systems that are very
similar to the known radio binaries do not exhibit radio emission.
The statistics of these results would be fairly well explained if
the typical radio binary spent only a few to several per cent of
the time in active flaring states. Radio variations in Algol have
been found to be classifiable into roughly three types. The most
easily recognizable is similar to that shown in Fig. 12, but
occasionally showing an apparent double peak at the highest levels.
The second type is a general, rather feature-less variability with an
essentially flat spectrum. The third type is the very rare occurrence
of non-thermal behavior. To get some idea of the relative occurrence

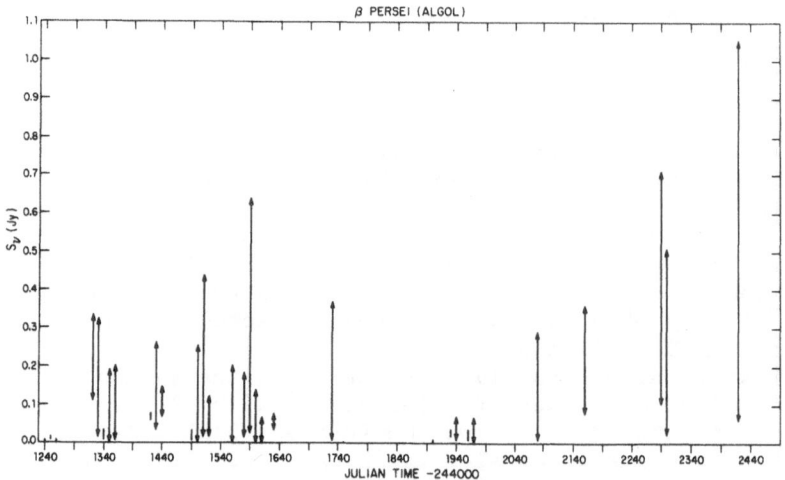

Fig. 11. The range of radio variations observed in Algol over a
four year period from Oct. 1971 to 1975 as measured at 8 GHz with
the NRAO interferometer.

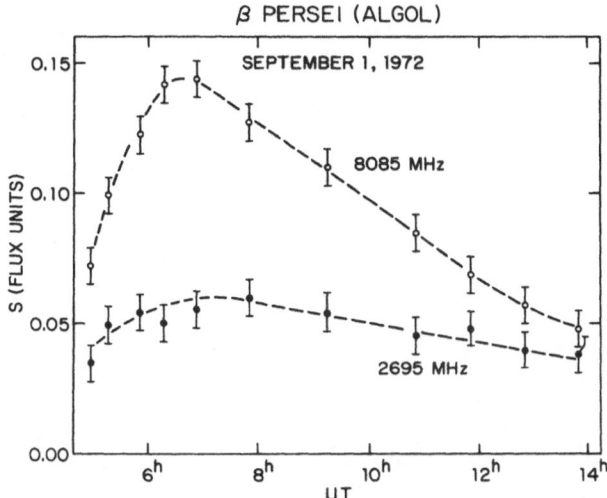

Fig. 12. A Sept. 1, 1972 radio event in Algol which is one of the few observed at 2695 and 8085 MHz which shows an essentially complete event.

of these types, Table IV is an attempt to describe the statistics of Algol "events", in a modified version of the results of Gibson (1975). The "flares" described in Table IV are all of the type where the flux was stronger at 8 GHz than at 2.7 GHz, with Fig. 12 being an example of the "typical" way such events evolve. The

Table IV

"Statistics" of Radio Events in Algol

Description of "Event"	Times Observed
flat spectrum but variable	56
single peaked flare	21
double peaked flare	11
flat spectrum evolving to beginning flare	3
flare decaying to flat spectrum	7
steady flat spectrum behavior	7
single point with flat spectrum	4
single point with "flare-like" spectrum	3
single point with non-thermal spectrum	1
definite non-thermal events	3
Total	116

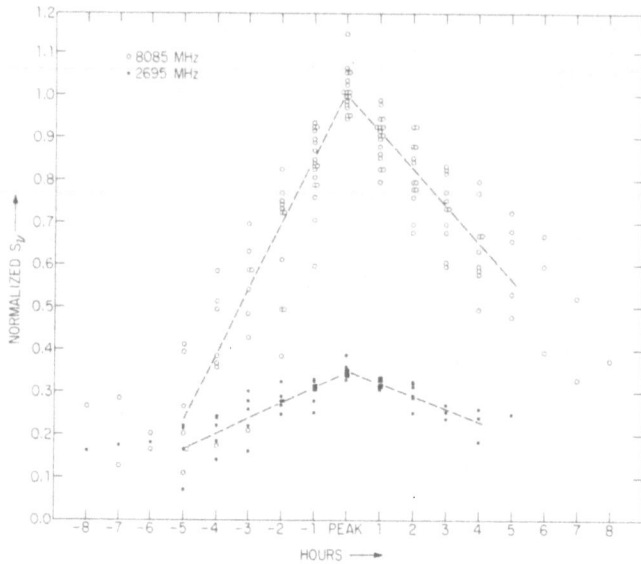

Fig. 13. A number of Algol peaking events at 2695 and 8085 MHz
are normalized with respect to the peak at each frequency and
plotted as a function of time before and after the peak.

three non-thermal events are of very special interest. On Jan. 23,
1975 a clearly decaying non-thermal event was followed by the
beginning of a more typical event dominant at higher frequencies.
On July 11, 1972, a uniquely short event, lasting only two hours
(Pooley and Ryle 1973), was observed to decay exceptionally rapidly
as a non-thermal source (Hjellming et al. 1972c). Finally, on Sept.
12, 1974, the only known example of a typical high-frequency dominated
flare evolved into a non-thermal decay (as seen in Fig. 10 for Ari)
(Gibson 1975).

       Gibson (1975) has used the available data on Algol events that
have clearly identifiable peaks to answer the obvious question
about how similar different Algol events might be. Fig. 13 shows
a number of things about the Algol peaking events. First of all,
their "shape" is quite similar with decay times somewhat longer than
rise times. Secondly, the major parameter distinguishing different
Algol peaking events is the flux density at the peak, or the integrated radio flux density if you note that the shape and duration
are roughly constant. Thirdly, the typical lifetime from initial
rise to final decay is roughly 16 hours as first pointed out by
Hjellming (1972).

       The most important radio observations of Algol in recent times
have been for the two events for which the actual size of the
evolving radio source was measured using the VLBI technique. The
first of these was a 0.7 Jy event on May 4-5, 1974 when Clark

et al. (1975) observed a slightly resolved radio source at 8 GHz
with a 20 million wavelength baseline. This established that the
angular size of the radio source was, at that time, about 0.004
arcsec, or 0.1 a.u., with an apparent brightness temperature of
4×10^8 K. The data were consistent with either a source expanding
at 500–1000 km/sec or a stationary elliptical source. A Jan. 16,
1975 event (Gibson et al. 1975) was observed at the 1 Jy level at
8 GHz by Clark et al. (1976) with 20, 85, and 100 million wavelength
baselines. The interpretation of this data is complex, but without
question the radio source was roughly 0.0018 arcsec in angular
size, corresponding to 0.045 a.u., with a brightness temperature of
10^{10} K.

The importance of the above results are obvious when one recalls
that the semi-major axis of the close double in the Algol triple
system is 0.066 a.u., and the diameters of the two stars are both
0.03 a.u. Therefore, for the two events where the angular size of
the radio source was measured, the radio source was definitely
larger than the size of either star, and was essentially the size
scale of the environment between the two stars. This rules out the
plasma wave model for radio star emission proposed by Jones and
Woolf (1973), because in their model the radio emitting region has
the size of the scale height of the stellar atmosphere.

Because of the very high brightness temperatures, the thermal
model for Algol radio emission initially discussed by Hjellming
(1972) is ruled out for the flaring events observed by VLBI. Whether
thermal emission is the origin of the very common, flat spectrum
radio emission cannot be clearly proven or disproven. The result
of all this is that the most fruitful working hypothesis for explaining the Algol radio emission, and presumably that of similar radio
binaries, is that of incoherent synchrotron radiation from relativistic electrons in a relatively dense plasma environment. The
latter results in extensive self-absorption and external absorption
of the radio emission and is the simplest explanation for the relative weakness of the lower radio frequencies in typical flare events.

8. WORKING HYPOTHESES AND PROSPECTS FOR FUTURE WORK

It is worth closing this review of our knowledge of radio stars by
emphasizing the working hypotheses that presently look fruitful and
the areas where future work is likely to be important.

On the theoretical side there are three prime areas that need
extensive work in connection with radio stars. The first and most
obvious is to further explore and make models based on synchrotron
source hypothesis. One needs to consider all possible energy loss
mechanisms and the special case of dense plasma conditions in the
radiating regions which probably plays an important role in both
radio emitting X-ray stars and the "normal" radio binaries.
Different geometrical situations should be considered: spherical,
double, and hemi-spherical. In addition, ram pressure confinement must

play a major role in the possible expansion of the radio sources
because of the relatively dense plasma environments. The second
area is the need to put stellar wind models, both static and
dynamic, on a sound basis, considering different energy supply
mechanisms, different geometries, different heating mechanisms, and
perhaps transient solutions. The work by Davidsen and Ostriker
(1974) on the importance of stellar winds for X-ray sources has had
considerable impact on the subject, and, as mentioned a number of
times, the need for maintaining relatively high magnetic field
strengths at considerable distances from stellar surface will frequently require invocation of some sort of stellar wind. Finally,
the ultimate and most important problem introduced by all the radio
observations of star and stellar systems is that of how to supply
and accelerate relativistic electrons in single and double star
environments. The energies involved in relativistic particles are
several orders of magnitude greater than we are familiar with on
the Sun.

On the observational side there are four main areas where
fruitful data may be obtained. First and foremost is the need for
what can be called broad-band radio spectroscopy, that is, high
sensitivity, simultaneous observations at many radio frequencies
in the 1 to 20 GHz range. Model building for stellar radio sources
will have very little information to constrain the models without
this type of data. Secondly, there is the need for VLBI "mapping" of
Cyg X-3, Algol, and the other radio stars. The ability to measure
changes in the structure of radio sources on short time scale is
unique to the radio star problem. Thirdly, there is an obvious extension of the radio spectroscopy already mentioned to coordinate radio,
optical, infra-red, and sometimes X-ray observations of radio stars.
Fourthly, there is the need for sensitive, high resolution mapping
of Sco X-1 and other radio stars that might have faint double
source companions; it would be very interesting to find the double
source phenomena extending all the way down to the stellar radio
sources.

REFERENCES

Ables, J.G., 1969, Ap. J. 155, L27.
Altenhoff, W.J., and Wendker, H.J., 1973, Nature, 241, 37.
Andrew, B.H., and Purton, C.R., 1969, Nature, 218, 855.
Andrews, A.D., 1969, I.A.U. Inform. Bull. Variable Stars, No. 325.
Becklin, E. E., Kristian, J., Neugebauer, G., Wynn-Williams, C.G.,
1972, Nature, 239, 130.
Becklin, E.E., Neugebauer, G., Hawkins, F.J., Mason, K.O., Sanford,
P.W., Matthews, K., Wynn-Williams, C.G., 1973, Nature, 245, 302.
Braes, L.L.E., and Miley, G.K., 1971a, Nature, 232, 246.
Braes, L.L.E., and Miley, G.K., 1971b, Astron. Ap., 14, 160.
Braes, L.L.E., and Miley, G.K., 1972, Nature, 237, 506.
Chu, K.W., and Bieging, J.H., 1973, Ap. J., 179, 22.

Clark, T.A., Hutton, L.K., Ma, C., Webster, W., Hinteregger, C.A., Knight, C.A., Rogers, A.E.E., Whitney, A.E., Shapiro, I.I., Wittels, J.J., Niell, A.E., and Resch, G.M., 1976, Ap. J. (in press).

Clark, B.G., Kellermann, K.I., and Shaffer, D., 1975, Ap. J. (Letters), 198, L123.

Cowley, A.P., and Crampton, D., 1975, Ap. J. (Letters), 201, L65.

Davidsen, A., and Ostriker, J.P., 1974, Ap. J., 189, 331.

Gibson, D.M., Viner, M.R., and Peterson, S.D., 1975, Ap. J.(Letters), 200, L43.

Gibson, D.M., 1975, Ph.D. Thesis, University of Virginia.

Gibson, D.M., and Hjellming, R.M., 1974, P.A.S.P., 86, 652.

Gibson, D.M., Hjellming, R.M., and Owen, F.N., 1975, Ap. J.(Letters), 200, L99.

Gibson, D.M., Hjellming, R.M., and Owen, F.N., 1976 (in preparation).

Gregory, P.C., Kronberg, P.P., Seaquist, E.R., Hughes, V.A., Woodsworth, A., Viner, M.R., Retalleck, D., 1972a, Nature, 239, 440.

Gregory, P.C., Kronberg, P.P., Seaquist, E.R., Hughes, V.A., Woodsworth, A., Viner, M.R., Retalleck, D., Hjellming, R.M., Balick, B., 1972b, 239, 114.

Gregory, P.C., and Seaquist, E.R., 1974a, Nature, 247, 532.

Gregory, P.C., and Seaquist, E.R., 1974b, Ap. J., 194, 715.

Herrero, V., Hjellming, R.M., and Wade, C.M., 1971, Ap. J.(Letters), 166, L19.

Hjellming, R.M., 1972, Nature Phy. Sci., 238, 52.

Hjellming, R.M., 1973, Ap. J.(Letters), 182, L29.

Hjellming, R.M., and Balick, B., 1972, Nature, 239, 443.

Hjellming, R.M., and Blankenship, L., 1973, Nature Phy. Sci., 243, 81.

Hjellming, R.M., Blankenship, L., and Brown, R.L., 1974, Ap. J. (Letters), 194, L13.

Hjellming, R.M., Gibson, D.M., and Owen, F.N., 1975, Nature, 256, 112.

Hjellming, R.M., and Wade, C.M., 1970, Ap. J.(Letters), 162, L1.

Hjellming, R.M., and Wade, C.M., 1971a, Ap. J.(Letters), 164, L1.

Hjellming, R.M., and Wade, C.M., 1971b, Ap. J.(Letters), 168, L21.

Hjellming, R.M., and Wade, C.M., 1971c, Ap. J.(Letters), 168, L115.

Hjellming, R.M., and Wade, C.M., 1971d, Science, 173, 1087.

Hjellming, R.M., and Wade, C.M., 1973, Nature, 242, 250.

Hjellming, R.M., Hermann, M., and Webster, E., 1972a, 237, 507.

Hjellming, R.M., Webster, E., and Balick, B., 1972b, Ap. J. (Letters), 178, L139.

Hjellming, R.M., Wade, C.M., and Webster, E., 1972c, Nature Phy. Sci., 236, 43.

Holt, S.S., Boldt, E.A., Kaluzienski, L.J., Serlemitsos, P.J., 1975, Nature, 256, 108.

Jones, T.W., and Woolf, N.J., 1973, Ap. J., 179, 869.

Kahn, F.D., 1974, Nature, 250, 124.

Kellermann, K.I., and Pauliny-Toth, I.I.K., 1966, Ap.J., 145, 953.

Kunkel, W.E., 1969, Nature, 222, 1129.

Lanque, R., Lequeux, J., and Nguyen-Quang Rieu, 1972, Nature Phy.

Sci., <u>239</u>, 119.
Lovell, G., 1969, Nature, <u>222</u>, 1126.
Marscher, A.P., and Brown, R.L., 1975, Ap. J., <u>200</u>, 719.
Mason, K.O., Becklin, E.E., Blankenship, L., Brown, R.L., Elias,
J., Hjellming, R.M., Matthews, K., Murdin, P.G., Neugebauer, G.,
Sanford, P.W., and Willner, S., 1976, Ap. J. (July 1 issue).
Parsignault, D.R., Gursky, H., Kellog, E.M., Matilsky, T., Murray,
S., Schrier, E., Tananbaum, H., and Giacconi, R., 1972, Nature
Phy. Sci.,<u>239</u>, 123.
Peterson, F.W., 1973, Nature, <u>242</u>, 173.
Pooley, G.G., and Ryle, M., 1973, Nature, <u>244</u>, 270.
Ramaty, R., Cheng, C.C., and Tsuruta, S., 1974, Ap. J., <u>187</u>, 61.
Sandage, A., Osmer, R., Giacconi, R., Gorenstein, P., Gursky, H.,
Waters, J., Bradt, H., Garmire, G., Sreekantan, B., Oda, M.,
Osawa, K., and Jugaku, J., 1966, Ap. J., <u>146</u>, 316.
Sanford, P.W., Ives, J.C., Bell Burnell, S.J., and Mason, K.O.,
1975, Nature, <u>256</u>, 109.
Seaquist, E.R., 1967, Ap.J., <u>148</u>, L23.
Seaquist, E.R., Gregory, P.C., Perley, R.A., Becker, R.H., Carlson,
J.B., Kundu, M.R., Bignell, R.C., and Dickel, J.R., 1974, Nature,
<u>251</u>, 394.
Tananbaum, H., Gursky, H., Kellog, E., Giacconi, R., and Jones, C.,
1972, Ap. J.(Letters, <u>177</u>, L5.
Wade, C.M., and Hjellming, R.M., 1971a, Nature, <u>235</u>, 271.
Wade, C.M., and Hjellming, R.M., 1971b, Ap. J.(Letters), <u>163</u>, L105.
Wade, C.M., and Hjellming, R.M., 1971c, Ap. J.(Letters), <u>163</u>, L65.
Wade, C.M., and Hjellming, R.M., 1971d, Ap. J., <u>170</u>, 523.
Wade, C.M., and Hjellming, R.M., 1972, Nature, <u>235</u>, 270.
Weymann, R., and Chapman, 1965, Ap.J., <u>142</u>, 1268.
Woodsworth, A., and Hughes, V.A., 1973, Nature Phy. Sci., <u>246</u>, 111.

X-RAY OBSERVATIONS

R. Giacconi

Center for Astrophysics, Harvard University,
Cambridge, Massachusetts, U.S.A.

1. EARLY HISTORY

The detection and study of cosmic sources of X-ray radiation has
progressed very rapidly since the discovery of the X-ray star,
Sco X-1, in 1962 (Giacconi et al. 1962). Rocket experiments in
the 1960's established the existence of X-ray emission from the
Crab Nebula (Bowyer et al. 1964b); identified the optical counter-
part of Sco X-1 and Cyg X-2 (Sandage et al. 1966; Giacconi et al.
1967); and detected the existence of the first extragalactic X-ray
source in the Virgo cluster of galaxies (Byram et al. 1966). Our
catalogue of X-ray sources had grown to about 30 sources by the
late 1960's. In addition, an intense diffuse background of X-ray
radiation, believed to be due at least in part to extragalactic
sources, had been detected since the first rocket flight and its
characteristics had been studied in later missions.

While considerable progress was thus being made by use of
short duration rocket flights, several important questions remained unanswered. For instance, while it was understood that in
Sco X-1 the X-ray emission corresponded to 10^3 times the luminosity
of our own Sun at all wavelengths, and exceeded the visible light
emission from the system by a factor of 10^3, the basic source of
this very large energy loss could not be determined. More significantly, due to the lack of positional accuracy and consequent
inability to identify optical or radio counterparts of a sufficiently
large sample of the X-ray sources, the connection between X-ray
observations and the main body of astrophysical knowledge relating
to stellar and galactic evolution could not be established.

The situation was considerably changed with the advent of the
first orbiting X-ray observatory "UHURU", launched on December 12,

G. Setti (ed.), The Physics of Non-Thermal Radio Sources, 229–269. All Rights Reserved.
Copyright © 1976 by D. Reidel Publishing Company, Dordrecht-Holland.

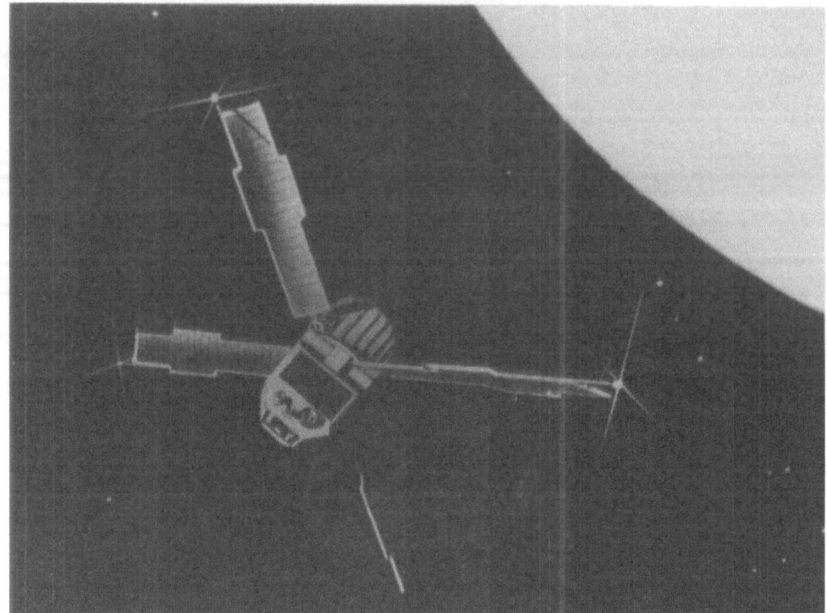

Fig. 1.   An artist's conception of the UHURU satellite, the first
orbiting X-ray observatory, launched on December 12, 1970, from
the San Marco platform in Kenya (from Giacconi, 1976).

1970, from the San Marco platform in Kenya and operated continuously
for four years (Fig. 1). The long time available for observations
(years, instead of the few minutes available in rocket flights), and
the much finer temporal and spatial resolution, have resulted in a
qualitative change in our understanding of the X-ray sky. The increased positional accuracy (~ 1 arc minute) was particularly important because it made possible the identification and study of the
optical and radio counterparts of X-ray sources. Several later satellite missions, such as OAO, OSO-7, ANS and UK 5, have confirmed
and extended the UHURU results. With these missions, we have reached a new level in understanding X-ray emission mechanisms and the
connection between X-ray sources and many of the astrophysical objects of greatest present interest such as collapsed objects (neutron stars and black holes), quasars and Seyfert galaxies, and
clusters of galaxies.

2. THE X-RAY SKY

The UHURU catalogue of X-ray sources (Giacconi et al. 1974) contains 161 sources which have been detected in the 2 to 6 keV energy
range. The weakest source in the survey corresponds to a flux of
about 3×10^{-11} erg cm^{-2} sec^{-1} at Earth, the strongest to about
3×10^{-7} erg cm^{-2} sec^{-1}. A map of the sky in galactic coordinates
is shown in Fig. 2. The sources divide naturally into two groups:

those that are clustered at low latitude along the galactic plane
(The Milky Way) and those that are spread more uniformly about the
sky. We believe that most of the sources we observe along the
Milky Way are stellar systems in our own galaxy. Most of those we

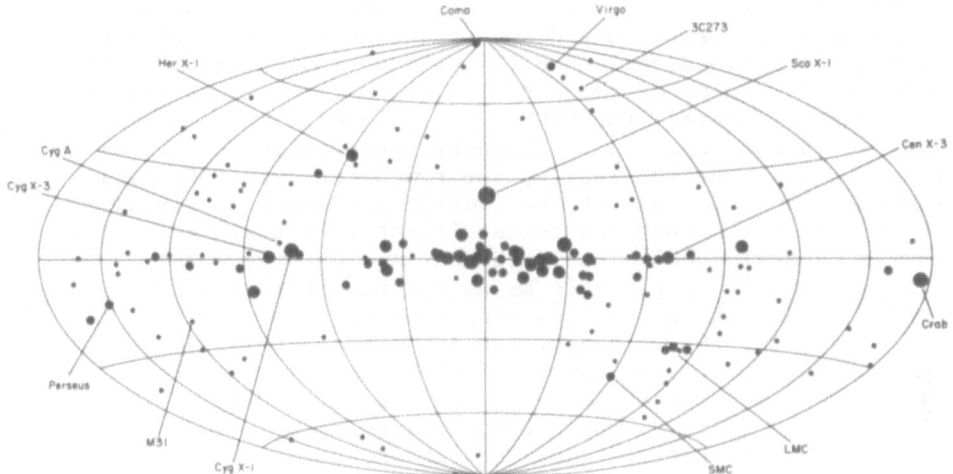

Fig. 2. A map of the X-ray sky in galactic coordinates (from
Giacconi, 1976).

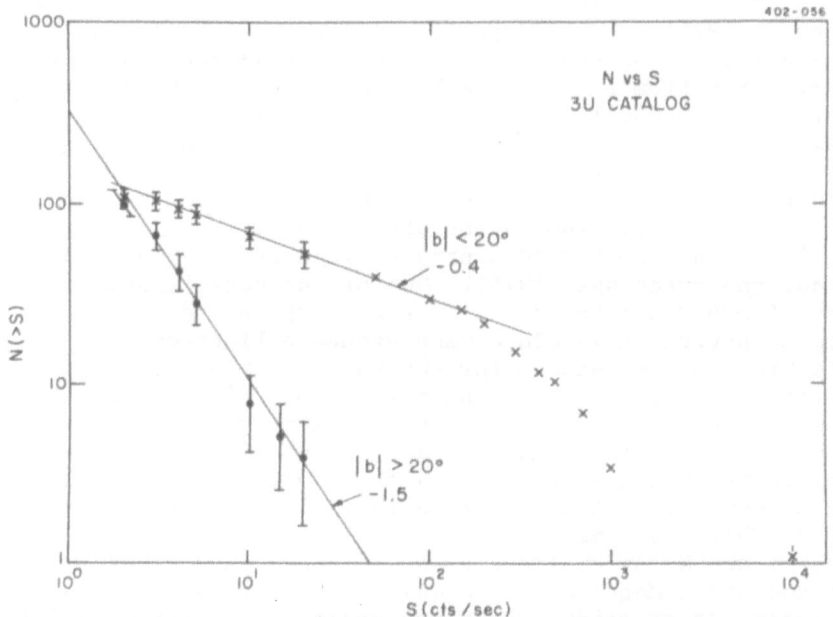

Fig. 3. The log N – log S relationship for low (|b| < 20°) and
high (|b| > 20°) galactic latitude X-ray sources.

observe at high galactic latitude (toward the galactic pole) are
connected with extragalactic objects. The qualitative difference
between the sources at low and high latitude is shown clearly if
we plot the integral number of sources N of intensity greater than
S. The low latitude sources show a weak dependence of N on S, as
expected for a disk or spiral arm population – the departure at
high energy being presumably due to a spread in the intrinsic
luminosity of close sources. High latitude sources show a steep
dependence of N on S, as expected for a spherical distribution.

Studying in detail the UHURU data regarding the distribution
of sources in the sky, the relationship between the number of
sources and apparent luminosity, the rough spectral information
available, and making use of the identifications from visible
light objects from UHURU and rocket flights, we could reach the
following general conclusions (Giacconi and Gursky, 1974):
(a) The luminosity of galactic X-ray sources is in the range
10^{35-38} erg/sec. X-ray sources are among the most luminous objects
in our galaxy and the mechanism for production of X-rays in these
objects must be therefore connected with their basic energetics.
(b) The number of sources in our galaxy in this range of luminosity
is of the order of 100. Thus, X-ray stars are only 10^{-9} times as
abundant as normal stars. (However, if one considers only the
class of massive blue supergiant stars, then about 1% are found
to be X-ray sources.)
(c) Of the identified galactic sources, 7 are supernova remnants,
and 6 are associated with binary stars. A recent review by Gorenstein and Tucker (1975) lists 7 supernova remnants, in addition
to the Crab Nebula, as the seat of X-ray emission in the range of
intrinsic luminosity from 2×10^{34} to 10^{37} observed in the range
of energy between 0.16 and 10 keV. The energy sources for X-ray
emission from supernova remnants are a) loss of rotational energy
by the pulsar in the case of Crab, and b) conversion of the kinetic
energy of ejecta from the supernova explosion by interaction with
the ambient interstellar medium in the other SNR's.

Table I lists the identified binary sources as of January 30,
1975 (Tananbaum and Hutchings, 1975). One of the most important
contributions of UHURU was to establish the energy source for these
X-ray sources as accretion in close mass exchange binaries.

All galactic sources, except for the SNR's, exhibit violent
time fluctuations with substantial changes in intensity in times
as short as 10^{-3} sec (Cyg X-1), but more prevalently in time spans
of seconds, up to years. For these reasons, the emission is believed to originate in small regions and the sources go under the
generic name of compact. It is possible that all compact sources
are members of binary systems.
(d) All galaxies are X-ray emitters at some level. Normal galaxies
(M 31, Large and Small Magellanic Clouds) emit about 10^{39} erg/sec
due to the integrated emission from X-ray stars. In these galaxies
$L_x/L_{rad} = 10^{-5}$ (L_x is the X-ray and L_{rad} the visible light luminosity) (Giacconi and Gursky, 1974).

Table I

Parameters of X-ray Binaries[†]

System	Primary Spectrum	Light Amplitude	X-ray Eclipse Duration	Estimated Inclination Angle	R/R_{crit}	q	M_P	M_X
CYG X-1 226868	O9.7 Iab	0.08	---	27°	0.98	1.6	22	14
SMC X-1 SK160	B0	0.09	±26°	70°	0.93	12	25	2
3U0900-40 77581	B0Ib	0.09	±35°	80°	0.84	17	45	2.6
3U1700-37 153919	O6f	0.06	±58°	90°	0.8	20	27	1.3
CEN X-3 KRZ STAR	O6f	0.08	±45°	90°	0.93	25	17	0.7
HER X-1 HZ HER	A7-B0	1.5 ON 0.3 OFF	±28°	90°	1.0 0.94	2.0	2.0	1.0
SCO X-1 —	IRREG.	0.25	---	LOW	~1?	0.5?	0.5?	1?
3U0352+30	B0e	---	---	90°	0.2	0.5	20?	40?

[†]Tananbaum and Hutchings, 1975.

(e) In active galaxies, such as Seyferts (NGC 4151), radio galaxies (NGC 5128), or quasars (3C 273), the X-ray luminosity is much greater, 10^{42} to 10^{46} erg/sec, and also L_x/L_{rad} is much greater. It is likely that the X-rays we observe originate in small regions near the nucleus of the galaxies where the occurrence of explosive events has been inferred from optical and radio studies.

(f) X-ray emission associated with clusters of galaxies appears to originate from extended low density regions between galaxies, as well as from individual galaxies. While there exists no obvious relation between X-ray emission and the extended radio and optical halos observed in clusters, the various emissions occur in over-lapping regions of space necessitating a single underlying pheno-menon. The luminosity of clusters is of 10^{43} to 10^{45} erg/sec. When an active galaxy is present, as in Virgo and Perseus, the X-ray emission appears centered on the active galaxy. At present, we attribute the extended X-ray emission to the existence of a hot intergalactic gas containing approximately the same mass as is present in the condensed galaxies of the cluster.

(g) The diffused X-ray background in the 2 to 6 keV range appears to be isotropic to within small observational limits (< 5%). This implies an extragalactic origin at large distances. If due to the contribution of unresolved single sources, these objects must be-long to a different class from the extragalactic sources already detected. If due to diffused mechanisms, then its origin must be tied to some aspect of the intergalactic medium. Its study might then have a bearing on a number of outstanding astronomical problems, such as the mass density of the Universe, early galaxy formation, and galactic activity.

We must examine critically some of these conclusions in light of recent reinterpretation of the data and of new results.

The major changes in the general picture expressed above have been as follows:

(1) On the basis of the published UHURU data, Margon and Ostriker (1973) had argued that the X-ray sources exhibited a narrow range of intrinsic X-ray luminosity from 10^{35} to $10^{38.8}$ erg sec^{-1}. They speculated that the upper limit of the luminosity was fixed by the Eddington limit ($L_{crit} \simeq 10^{38}$ M/M$_\odot$ erg sec^{-1}) and that plausible mechanisms for X-ray generation (such as self-excited models) could be invoked which would tend to drive the X-ray emission right up to the limit.

This view must be modified due to recent results of further analysis of the UHURU satellite by Forman, Jones and Tananbaum (1976) which show the existence of numerous sources at low galactic latitude $|b^{II}| < 5°$. These sources exhibit a wide distribution of galactic longitude and fit the extrapolation of the ln N – ln S distribution for sources at low galactic latitude.

Thus, they do not appear to represent a new class of galactic sources, but rather simply less luminous members of the same class or classes of sources already observed. This could be readily understood if one assumed that all sources are due to accretion in

close mass exchange binaries and took into account the very wide
range of masses and accretion rates that could be represented.

The view that large numbers of yet undetected low luminosity
sources could be present has been strengthened by the discovery
by SAS-3 of X-ray emission from HZ 43 (Hearn and Richardson, 1975),
a white dwarf with an M dwarf companion and with intrinsic luminosity
of order of 10^{31} erg sec^{-1} with a spectrum consistent with black
body radiation from the surface of a white dwarf at about 10^5°K.
The energy powering the X-ray emission is likely in this case to
be due either to cooling of the surface of the white dwarf or to
nuclear burning rather than direct release of gravitation free-
fall energy from an accreting gas.

The discovery of X-ray emission from Capella and Sirius (Catura and Acton, 1975; Mewe et al. 1975) due presumably to coronal
emission, at 3 x 10^{29} and 10^{28} erg/sec respectively show further
that the view of the sky which emerged from UHURU was incomplete.
When discussing galactic X-ray sources, we must consider now
several classes of object with different intrinsic luminosity,
emission mechanisms and energy source spanning the range from
10^{28} erg sec^{-1} from the coronas of stars (including our own Sun)
to the 10^{39} erg sec^{-1} of SMC X-1.

It is, however, important to point out that the total integrated luminosity from all the weaker sources does not substantially
contribute to the total emission from the galaxy or to the diffused
soft X-ray background.
(2) The discovery of numerous transient X-ray events primarily by
the Ariel satellite, but also by OSO-7, UHURU, SAS-3 and ANS, both
along the galactic equator and at high galactic latitude raises the
question of whether these sources are qualitatively different from
the compact X-ray sources in the UHURU survey.

Willmore in a recent review (1976) reports the detection of
sources with lifetimes of order of days to weeks with a frequency
of about 100/year and intensity greater than 100 UHURU cts/sec.

Typically they reach for a time at least the intensity of
Crab and last between a week and twenty weeks.

The galactic distribution of seven such events from Willmore,
1976, clearly shows that the events are prevalently concentrated
along the galactic plane. Intrinsic luminosity are estimated by
various means to be close to 10^{38} erg sec^{-1} at peak. An important
clue to their nature was the discovery that 2 of them exhibited
pulsations similar to those found in other binaries, such as Her
X-1 and Cen X-3, but of much longer duration (6.55 ± 0.010 minutes
for A1118-61 and 1.737 ± .003 minutes for A0535+26). These periods
were at first puzzling because, if due to rotation of a neutron
star accreting from a companion, the neutron star had to be considerably slowed down. If due to an orbital period, such a short
period implies 2 compact sources; for instance, a neutron star and
a white dwarf, - it is difficult to imagine matter loss mechanisms
from the white dwarf. The situation was somewhat clarified by the
discovery of slow pulsations (283 seconds) in 3U0900-40, a well-

studied binary which is believed to consist of a 1.4 $M_\odot$ neutron
star with a 20 solar mass companion in eccentric orbit. This
observation established a link between transient X-ray sources of
this type and "normal" X-ray binaries and gave credence to the
models invoking a variation in the matter accretion rate to explain
the flux increases. The recent survey by Holt (1975) and by Forman,
Jones and Tananbaum (1976) shows that the frequency of such events
must be of order of only 10/year.

Thus, transient sources of this type do not seem to require
a different class of objects and their integrated luminosity is
small compared to the total galactic luminosity.

The work by Forman, Jones and Tananbaum (1976) further shows
that the transient low latitude sources are only an extreme example
of a rather common characteristic of all compact sources, namely
extended low and high emission periods. In particular, the detection of AO535+26 more than four years prior to its transient
brightening phase seems to indicate a behavior quite similar to
that found in 3U1630-47, 3U1901+03, 3U0115+63 and 3U1735-28, namely,
long period of extremely low intensity alternating with outburst
type events.

(3) A few high latitude transient sources have been observed by
Pounds (1975), Rappaport (1976) and by Ulmer and Murray (1976).
This is particularly interesting because of the implications with
respect to the nature of the unidentified high latitude sources.
If we examine the Log N – Log S distribution for objects of $|b^{II}|$ >
> 20° we find that most of the weak sources between 10 and 2 cts
per second are not identified. Within the small intensity range
available to us they follow an integral distribution of $S^{-3/2}$ as
would be expected either for an extragalactic of distant sources,
or a population of galactic sources quite close to us. The nature
of these sources, if extragalactic, is one of the more fascinating
and still unresolved puzzles of X-ray astronomy. If they are
extragalactic, then their ratio of X-ray to total luminosity is
larger than for any known extragalactic object, including 3C273,
and if the distribution extended unaltered to smaller intensity,
they would provide a substantial fraction of the isotropic background.

Up to now, only tentative identifications have been made in one
case with a QSO (3U0138-01=NAB0137) in two cases with clusters of galaxies (3U1555+27=A2142, 3U0405+10=A478), both therefore with extragalactic objects (Bahcall et al. 1975; Schreier et al. 1975; Rowan-
Robinson and Fabian, 1975; Giacconi et al. 1974; Elvis et al. 1975).
If we were to observe short time variability, this might indicate a
scale too small to be compatible with emitters in external galaxies.
The Pounds' reported event and those reported by Markert and Ulmer
are inconclusive since their variation occurs over days or longer and
are compatible with the variations observed in the nucleus of Cen A.
In fact, in the case of Pound, a tentative identification is proposed
with a B Lac object. The Rappaport event, however, seems to exhibit
characteristics incompatible with an extragalactic origin. The total

length of the event (<2000 sec) and its characteristics, including
the absence of a conspicuous extragalactic candidate are believed to
be suggestive of a galactic origin. If so, the authors suggest that
the object might be an example of accretion onto a white dwarf in
a binary system.

Do most high latitude unidentified sources exhibit this behavior? The answer is clearly no. In a recent study by Murray and
Ulmer it was shown that in a sample of 22 of these objects constant
emission was observed at intervals of years from most of them. The
tentative conclusion is therefore that most of the unidentified high
latitude sources are not transient, leaving the puzzle of their
identity still unresolved. It must also be concluded, however, that
a fraction of these events may well be of the type observed by Rappaport and, therefore, probably of galactic origin.
(4) The renewed interest in the X-ray sources in globular clusters
brought about by both theoretical considerations, as well as new
data from SAS-3 and ANS, makes them one of the most exciting current
topics in X-ray astronomy.

Globular clusters are believed to be almost as ancient as the
Galaxy itself (10 billion years), and in the most generally accepted
theory they are believed to have condensed at about the same time
that the galaxy itself was condensing from a gas cloud. This view
is supported by the fact that examination of the Hertzsprung-Russell
diagram for stars in the globular clusters demonstrates that most
of the stars are very old and are in intermediate or advanced evolution. Only very low mass stars, where time for evolution is very
long (up to a trillion years) are still found in the main sequence.
All others with greater masses have evolved long ago off the main
sequence track.

In the study of the binary X-ray sources we have found that
most such systems contain, in addition to a compact object, a young
massive supergiant having evolution times of the order of 10^8 years
at most. The discovery that several X-ray sources are associated
with globular clusters, where such young stars are not believed
to exist, presents us therefore with a considerable puzzle.

The UHURU survey identified 4 globular clusters as possible
optical counterparts of X-ray sources. Three of these identifications (NGC 6624, NGC 6441 and NGC 7078, or M 15) were confirmed
by the OSO-7 survey which also discovered two more cluster X-ray
sources. The sources were found to have intrinsic luminosity from
4×10^{36} to 7×10^{37} erg/sec and to exhibit intensity variations.
They are, therefore, similar in many characteristics to the galactic
binary sources. However, we immediately are faced with a problem:
we have 5 X-ray sources associated with the 120 known clusters.
This is a 100 times greater abundance of X-ray sources with respect
to stars for the clusters than for the galaxy as a whole. This
argument, which was first pointed out by Gursky and Schreier in
1974, therefore suggests that globular cluster X-ray sources are
qualitatively different from the binary X-ray sources found elsewhere in the galaxy. If one attempts to explain the globular

clusters X-ray sources as the product of the evolution of stars
which were born as binaries, we are confronted with the basic
problem that any massive binary system which could have evolved
so as to form a neutron star or a black hole could not still contain a massive nuclear burning companion providing the mass for
the accretion process. Such a companion would have burned out long
ago leaving the compact object without a source of accreting material.

No doubt, therefore, that we must find a different explanation
about either the evolution or the nature of these sources. Several
possible explanations are currently being tested observationally.
The first invokes the recent formation of binary systems by capture. The collapsed remnant of early massive stars which evolved
long ago must still be present in the clusters. Clark (1975)
suggested that these remnants can capture field stars which then
supply them with accreting material. Several others have advanced
specific suggestions as to how such capture could take place: the
main difficulty being in accurately assessing the frequency of such
processes. No actual observational evidence can yet be invoked

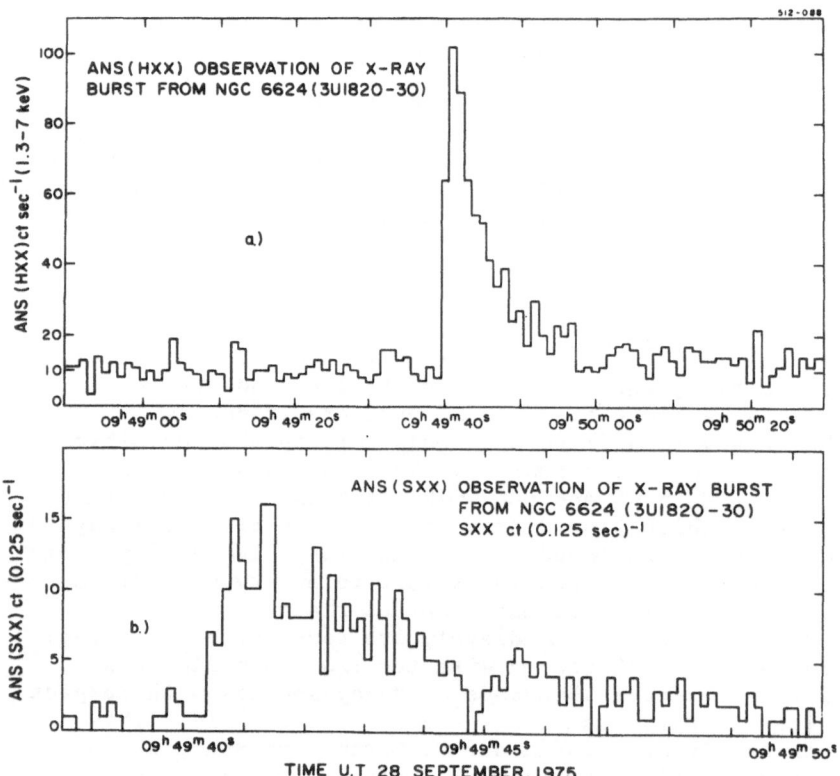

Fig. 4. Intensity profiles as seen by a) the hard X-ray experiment
(∿1 - 30 keV) and b) the soft X-ray experiment (1 - 7 keV).

to support this view. None of the globular cluster sources has
been shown to exhibit eclipsing behavior or regular pulsations
with Doppler shifts in frequency that could be considered actual
proof of the binary nature of the sources. Bahcall and Ostriker,
and Silk and Arons suggested as an alternative hypothesis that
X-ray sources in globular clusters are due to accretion onto black
holes of 10^3 solar masses. Peebles, in 1972, had pointed out that
superdense cores could be formed in the evolution of clusters and
proposed that such massive black holes would be the expected result
of the collapse of the central core. X-ray emission could, in the
view of the proponents of this theory, occur in those clusters
with deep enough potential wells (or high enough escape velocities)
which would be sufficient to retain in the cluster the gas emitted
from stars during successive traversals of the galactic plane. In
most clusters the escape velocity is such that the gas would be
lost as it is generated and therefore they could not be X-ray
sources. There are difficulties connected with this model as well,
since it is not obvious what are the measurable cluster conditions
leading to X-ray source formation in some, but not others, and also
why such sources, with black holes of 10^3 M_o, should be limited in
emission to 10^{38} erg/sec, the Eddington limit for stars of one
solar mass.

The recent discovery of X-ray bursts connected with globular
clusters (Grindlay et al. 1976; Clark et al. 1975) has opened an
entire new approach to their study. Fig. 4 shows the appearance
of the first X-ray burst discovered by Grindlay et al. (1976) from
the ANS satellite. The source NGC 6624 appears quiescent at a
luminosity equivalent to 10^{37} erg/sec and then in 1/2 second
rises to about 30 times the quiescent level, only to decay exponentially in 10 seconds to the previous level. Grindlay and
collaborators found two such bursts from the same source with
identical appearance. (This characteristic was confirmed by the
observation of 9 additional bursts by SAS-C.) They reasoned that
the same shape could not occur at random, but either reflected
some characteristic of the underlying source, or could be due to
some characteristic feature of the medium about the source which
would modify the appearance and spectrum of these bursts and give
them their characteristic features. They were struck in particular by the similarity of the appearance of these bursts with the
light curves observed in supernovae: namely, a fast rise, a peak
and then an exponential decay. Also, they noted that an essential
feature of the bursts was the hardening of the spectra as the
burst progressed. By analogy with the Morrison and Sartori models
with supernova events, they suggested that the most "natural" explanation for the events was as follows. An initial prompt pulse
was generated at the source. Some of the radiation would reach
the observer unmodified by the surrounding medium. Some could be
scattered by the surrounding medium and reach the observer with
different delays from different regions. This would result in an
exponential decay of the pulse. In order to explain the hardening

of the spectrum, they assume that the cloud is composed of a hot
gas. The scattering of the X-rays from the energetic electrons
in the cloud would then modify the spectrum as observed. The
characteristic exponential decay time gives then a measure of the
cloud dimensions, its temperature is known from the spectral measurement and they can then derive the value of the mass of the
central object necessary to keep this hot cloud bound to it at
that distance. The result is that the central object should be
of the order of 1000 M_o. The bursts of NGC 1820-30 would be then
the scaled up version of Cyg X-1. The slower rise time is, precisely as expected, due to the difference in mass. More recently,
Clark and his colleagues at MIT, found evidence that the pulses
from this cluster came in quasi-periodic fashion at intervals of
approximately 0.18 days persisting for at least 10 pulsations.
They also found additional bursts from other sources with different
characteristics and behavior but apparently denoted by the fact
that the peculiar features of the bursts from each source remain
constant and are, in effect, a characteristic thumbprint of that
source.

Only a week ago, re-examination of UHURU data by the Formans
has revealed a similar event from yet another cluster source but
with characteristic duration of decay time of ~25 sec (Jones and
Forman, 1976). This complex phenomenology then reveals a rich
new field of exploration in X-ray bursts phenomena. The relation
of these bursts to the gamma-ray bursts found by Strong and his
collaborators in 1973 is not clear. They appear somewhat different
in temporal and spectral characteristics. The lack of precise
location measurement for the gamma-ray bursts also makes it difficult to associate them with individual known sources, although it
is suggestive that the only tentative identification proposed by
Strong et al. 1973 is with Cyg X-1.

What all this means is rather obscure at the moment. The
general excitement, however, is quite understandable: such precise
and distinct behavior as observed in the bursting X-ray sources
had not previously been observed in any of the well-studied binary
X-ray sources. The energy involved, 10^{39} erg/sec, is so large that
this behavior must be tied at some fundamental level to the intrinsic energetics of the source. We have hopes that these studies
might shed light on the existence of massive black holes, and permit determinations of their parameters.

It should be clear from the above that while the fundamental
picture of the X-ray sky obtained by UHURU in 1970-72 still stands,
further studies have revealed a much greater richness and variety
of conditions and objects than we had then understood. The further
progress in X-ray astronomy which will be brought about in the next
two or three years by the flight of several NASA missions will
further expand the realm of contributions that X-ray astronomy can
make to the study of high energy processes in the universe.

3. ORIGIN OF THE ENERGY DISSIPATED IN X-RAY SOURCES

The large intrinsic luminosity characteristic of the brightest
of the X-ray sources in our own galaxy, as well as in external
galaxies, poses interesting questions regarding: the source of the
energy dissipated, the process by which this energy is expended
in the creation of high energy particles, or in the heating of
plasmas, and finally, on the details of the X-ray emission mechanism
itself.
    The first set of problems, namely the ultimate source of the
energy dissipation which we observe is the most fundamental question
we can ask about any system and its resolution has led to many of
the recent advances in modern astrophysics.
    When astronomers at the beginning of this century began asking
in earnest questions about the energy source for normal stars, including the Sun, the solution involved the opening up of a new
branch of physics, namely, the study of nuclear processes.
    We shall see that for many of the X-ray sources we discussed
in the introductory sections the energy source is not nuclear, but
still, involving as it does the release of gravitational or rotational energy, does not appear to require new physics. For some
sources such as the nuclei of active galaxies, the quasars, and
possibly for all those sources in which a black hole candidate is
proposed, the question is still open.
    I mentioned that for the majority of the bright galactic X-ray
sources the ultimate energy source appears to be the release of
gravitational or kinetic energy. For these sources, assuming we
are correct about the ultimate reservoir of energy, the problem
confronting us is the explanation of how this energy is efficiently
converted to produce high energy particles or high temperature
plasmas. In most cases these conversion mechanisms are not fully
understood. The study of possible emission mechanisms of the
X-rays permits us to relate the observed characteristic of the
radiation to the physical conditions at the source and thereby
to gain understanding of the conversion process.
    I will attempt to outline the current approaches to these
problems by freely mixing general considerations with descriptions
of the phenomenology of a few specific X-ray sources and their
interpretation.
    The only two sources of energy available for dissipation by
celestial bodies are: nuclear energy and gravitational energy. Normal
stars utilize nuclear energy derived from the synthesis of light
into heavy elements. Burning of atomic hydrogen into helium
liberates of order of 0.7% of the rest mass. Synthesis of heavier
elements delivers energy of the same order of magnitude. Roughly,
therefore, 1% of the rest mass of matter can be converted into
radiation by nuclear burning.
    In the case of 1 solar mass star, the rest energy is $M_\odot c^2 =$

= 2 x 10^{54} erg. Therefore, the available energy from nuclear
burning is of order of 2 x 10^{52} erg. This energy can only be
liberated when the temperature of the matter is sufficiently high
to permit ignition. Therefore, most stars like the Sun burn nuclear fuel at their core. The radiation produced in the interior
is degraded in energy as it proceeds to the outer layer until it
reaches the surface at the relatively low temperature of order of
10^4 K. The peak emission of a black body at this temperature is in
the visible. Most stars, therefore, emit most of their radiation in
the UV to IR range. In order to produce X-rays, temperatures of
millions of degrees are necessary and these temperatures can only
be reached by collective phenomenas (for instance, sound waves)
capable of attaining much greater temperatures, or of accelerating
high energy particles. Such phenomena are believed to be occurring,
for example, in the solar corona. It should be noted that since
they require a large decrease of local entropy, they are extremely
inefficient and can involve only a minute fraction of the total
radiated energy. Gold has pointed out that this may explain why
normal stars are such inefficient X-ray emitters ($L_x/L_{tot} \lesssim 10^{-6}$).

It is clear that such gentle, steady, low temperature heating,
as provided by nuclear sources, cannot, therefore, provide the
source of the high temperature, high luminosity energy releases we
encounter, for example, in supernova events, or in the pulsar radiation.

The other source of energy we have at our disposal is gravitational energy or kinetic energy. The negative gravitational
energy of a spherical object of mass M and radius R is of order
of GM^2/R, the precise value being a function of the density distribution. For the Sun this energy is of order of 10^{48} erg. Thus,
for most stars, including the Sun, the energy liberated during
contraction to its present state is negligible with respect to
the energy liberated in nuclear reactions.

However, if the star is allowed to contract, the negative
gravitational energy increases as 1/R. In white dwarfs (M = $M_\odot$,
R = 10^9 cm) the gravitational energy that can be liberated is of
order of 10^{50} ergs. For neutron stars (M = $M_\odot$, R = 10^6 cm) it is
of order 10^{53} ergs. Thus, the contraction of a star, which will
occur at the end of its nuclear burning phase, when the radiation
pressure of the interior can no longer sustain the outer layers
against gravity, can give rise to the release of gravitational
energy which, in the case of neutron stars, is of order 10% of the
rest mass energy.

How is the energy, produced by this contraction, released?
During the formation of neutron stars the supernova explosion expels
a shell that can carry away a significant fraction of the available
energy. A significant fraction of the energy can, however, remain
stored in the neutron star at formation, in the form of rotational
energy. If the parent star had a finite angular momentum, although
a fraction of this angular momentum may be dissipated by mass loss
and other effects, conservation of the residual angular momentum

will force an increase of angular velocity Ω of the star due to
the decreased moment of inertia. This will increase the rotational kinetic energy of the system. The maximum value for Ω
being given by setting the centrifugal force equal to the gravitational force at the equator: $GM/R^2 = R\Omega^2$, from which one can
derive that $W_{rot} = -1/3E_{grav}$ if ρ = constant. Thus, the rotational energy of the neutron star could store a significant fraction of the gravitational energy liberated in the collapse of order
of 10^{52} erg.

The significant fact about this reservoir of energy is that
the energy is concentrated in a single degree of freedom, which
(as pointed out by Rossi (1974) whose treatment of this subject
I have followed) yields an extremely low entropy condition. From
the thermodynamical point of view, there is no difficulty, therefore, to convert this energy, quite efficiently, in any other form
including the generation of high energy particles. How, in detail,
can this energy be transformed is not entirely clear even when we
believe that we have evidence that this process is occurring, as
we believe is the case for the NP 0531 in the Crab Nebula.

For many years since the early studies of Hubble in 1928,
who proposed that the Crab Nebula was the gaseous remnant of a
supernova explosion of 1054, attempts were made to explain the
intensity and shape of the continuous spectrum emitted by the
Crab. All such attempts based, as they were, on the assumption of
thermal processes, failed. The discovery in 1949 that Crab was
also a radio emitter of great power which could not be explained
as a thermal process increased the interpretative difficulties.
The solution was proposed by Shklovsky in 1953, who suggested that
the continuum spectrum both invisible and radio could be due to
synchrotron emission from high energy electrons spiraling in the
magnetic field of the nebula. The observation of polarization of
the optical continuum confirmed this view. While this suggestion
was extremely powerful in resolving the mechanism of radiative
emission, it left the problem of the energy source unsolved. The
discovery in 1963 that the Crab Nebula was an X-ray source further
complicated the problem (Bowyer et al. 1964). Electrons of 10^{12}
eV in a field of 10^{-3} to 10^{-4} gauss, as believed to exist in the
Nebula, produce optical photons. They have lifetimes to energy
loss of order of 10^3 years. Thus, one could conceive that the
electrons producing the optical synchrotron today had been accelerated in some mysterious way in the original explosion. However,
electrons of 10^{14} eV energy required to produce X-rays in a magnetic field of $10^{-3} - 10^{-4}$ gauss have much smaller lifetimes,
$\tau(\text{sec}) \simeq 4 \times 10^{14}/E(\text{eV})B^2(\text{gauss})$. Electrons of the energy necessary
to produce 10^5 eV X-rays could live only 5 years. Thus, one had to
conclude that an accelerating mechanism was now present in Crab
which could replenish the energy of the electrons dissipated through
radiation. This energy had to provide of order 10^{37} erg/sec. A
plausible explanation for this mysterious source came only with
the discovery of the Crab Nebula pulsar in 1968. The observations

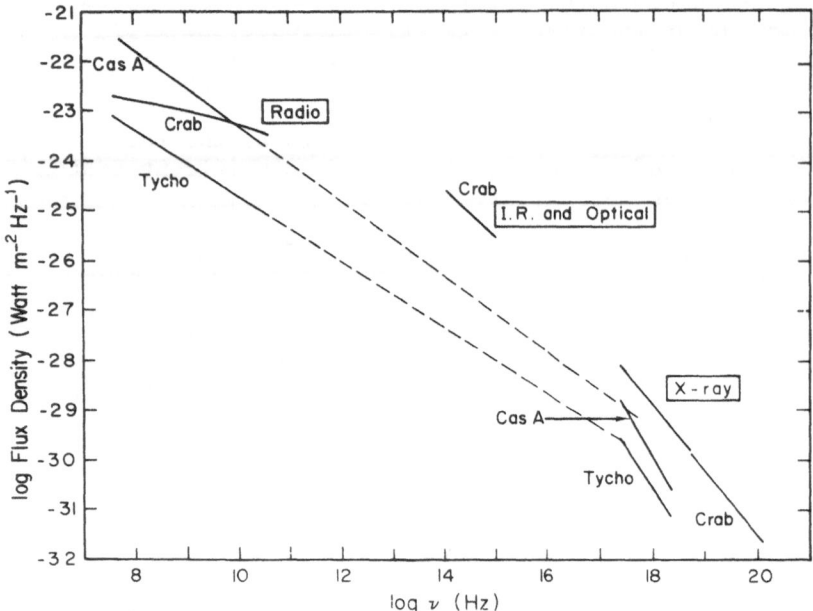

Fig. 5. The electromagnetic spectrum of the Crab Nebula, Cas A
and Tycho SN remnants.

show that pulsed emission from the Crab extends from radio waves
to gamma-ray energies (Grindlay et al. 1976) of 10^{11} eV (Fig. 5).
The pulsations occur with a period of 33 milliseconds (Fig. 6) which
is interpreted as due to the rotation of the neutron star producing
a fan beam of radiation. This period has been observed to undergo
a secular increase of 4 parts in 10,000 per year. If we adopt the
generally accepted view that the Crab pulsar is a neutron star of
about 1 M_O and about 10 km radius, we can compute its angular momentum to be of $\sim 10^{49}$ erg. A change in the angular velocity Ω
will therefore yield a loss of rotational kinetic energy W_r given
by

$$\dot{W}_r = I \Omega \dot{\Omega} \tag{3.1}$$

where I is the moment of inertia of order I = 5 x 10^{44} erg s^{-1}.
The observed change in Ω corresponds to an energy loss of $\dot{W}_r$ =
= 3 x 10^{38} erg s^{-1}. Thus, today it is generally believed that we
have found the source of energy both for the pulsar emission itself
and for the high energy electrons that stream in the nebula and produce by synchrotron the observed continuum emission. Of the problems we had set ourselves, namely the source of energy, the mode
of conversion and the mechanism for emission, we seem therefore to
have resolved the first and last. What about the mode of conversion?

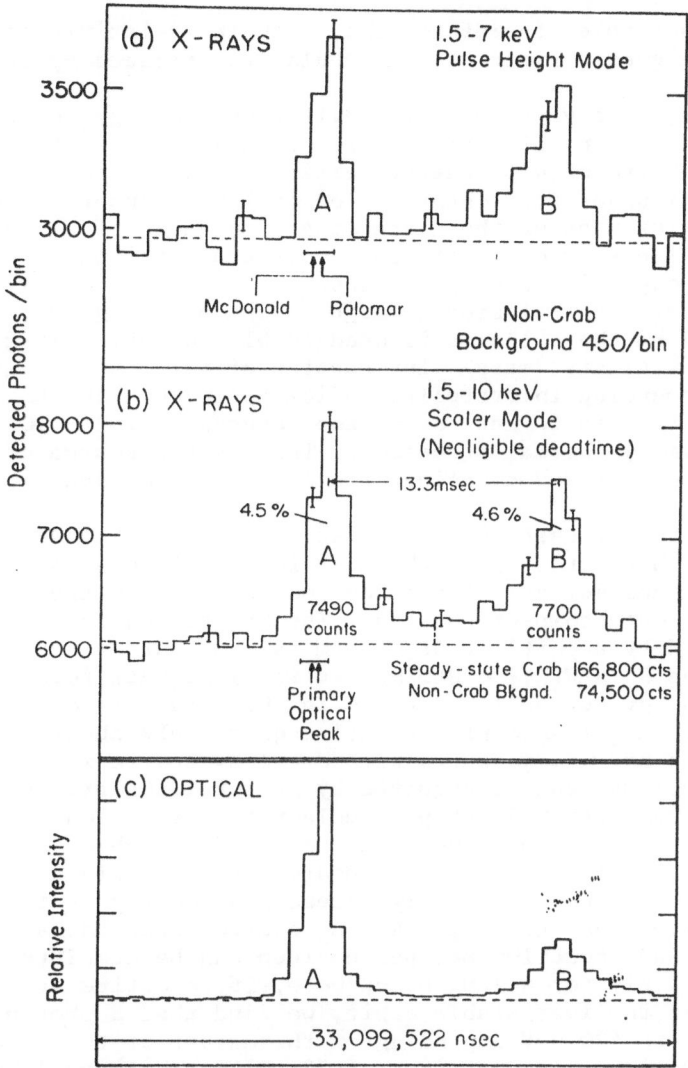

Fig. 6. Simultaneous optical and X-ray observations of the time structure of the Crab pulsar.

That is still poorly understood, but it is believed to be related to processes occurring in a rapidly varying magnetic field. From simple considerations it can be shown that dipole radiation from a rotating neutron star with a field 10^{12} gauss can generate large electric fields which can in principle accelerate particles to energies as large as 10^{16} eV. In real life the magnetic field will not be a dipole since the emitted plasmas will tend to stretch the field radially out. It is quite possible, therefore, that both the relativistic electrons and the magnetic field of the Nebula are produced by the pulsar.

From the above, we have seen that gravitational energy released from collapse is stored in rotational energy of the neutron star and released in the form of high energy particles and photons through the not well understood intervention of large magnetic fields. The question arises of whether there are other instances of X-ray emitters which utilize stored kinetic energy or gravitational energy possibly through different processes.

During the supernova explosion a large amount of the gravitation energy freed in the collapse is used in blowing off a large fraction of the initial star mass. Interaction of this expanding shell with the surrounding interstellar medium produces a shock capable of heating the gas to very high temperatures ($\sim 10^6 - 10^7$ K) which emit X-rays by thermal processes. This is the source of the energy for the soft X-ray emission from all old supernova remnants.

Yet another process, based on the release of gravitational energy, is believed however to play the most important role in the X-ray emission from compact galactic X-ray sources, namely the accretion of matter onto compact objects. The kinetic energy that can be acquired by a particle of mass m falling onto a star of mass M and radius R is $W_{fall} = GMm/R$. It is clear that for normal stars the energy per nucleon gained in the infall is quite small: for $M = 1 M_\odot$, $R = 1 R_\odot$ a proton will acquire only about 100 eV energy. Even for white dwarfs with $M = 1 M_\odot$ and radius 10^9 cm only about 0.1 MeV can be acquired by an infalling proton. This is to be compared with ~ 7 MeV per nucleon that is released in hydrogen burning. It is only when the radius of the object becomes quite small, as in the case of a neutron star or black hole, that the free-fall energy greatly exceeds the energy that can be liberated by nuclear burning. A 1 $M_\odot$ neutron star has a radius of $\sim 10^6$ cm and about 100 MeV per nucleon can be acquired in free fall. Similarly for a 1 $M_\odot$ black hole, if we define as radius the radius of the last stable orbit, we find that a proton can acquire from 60 to 400 MeV depending on the nature of the black hole (Lamb, 1974; Davidson and Ostriker, 1973; Blumenthal and Tucker, 1974; Giacconi, 1975a; and Giacconi and Gursky, 1974). If we consider the total accreted mass per second ($\dot{M}$) then the total energy produced is $dE/dt = GM\dot{M}/R$. The luminosity of the object depends on the efficiency with which the infall energy is converted in protons. Thus,

$$L = \alpha \ 10^{33}\left(\frac{M}{M_\odot}\right) \ \left(\frac{R}{R_\odot}\right)^{-1} \left(\frac{\dot{M}}{10^{-8}M_\odot/yr}\right) \text{erg s}^{-1} \qquad (3.2)$$

in which α is always close to 1.

The temperature T produced by thermalization of the flow is

$$T \sim \beta \, m_p \, \frac{GM}{kR} \sim 10^7 \beta \left(\frac{M}{M_\odot} \right) \left(\frac{R}{R_\odot} \right)^{-1} \tag{3.4}$$

where β ranges from 10^{-6} to 10^{-1} depending on how the gas is actually heated since the ultimate temperature attained will depend
on the balance between radiative losses and heating, and k is the
Boltzmann's constant.

From the previous formula, we see that even adopting for β
the value 10^{-6} for a 1 $M_\odot$ neutron star the temperature will still
be of order of 10^7 K. It is clear, therefore, that these simple
considerations show that provided a sufficient rate of accretion
($M \gtrsim 10^{-8} M_\odot$ year^{-1}) neutron star and black holes can radiate of
order of $L_x \sim 10^{38}$ erg s^{-1} at temperatures $> 10^7$ K as required
to produce X-ray photons.

As early as 1964 Zel'dovich and Guseynov, and Hayakawa and
Matsuoko, had suggested that this process may actually be occurring.
They had considered, however, isolated stars gathering matter from
the very tenuous interstellar gas ($\rho \sim 1$ particle cm^{-3}). In these
conditions the mass accretion rates to be expected fall short by
orders of magnitude in providing sufficient matter accretion to
power the X-ray sources we observe. In a close binary system, however, compact stars could efficiently gather material blown off by
a normal companion. Since the rate of matter loss during particular stages of stellar evolution can be as large as $10^{-3} M_\odot$ per year,
clearly we should have immediately thought of binary systems (in
which about 1/2 of all stars are found) as the possible source of
the observed X-rays. However clear the argument appears now, with
the advantage of hindsight, when it was first proposed by Shklovsky
in 1967 it did not receive general acceptance. This was due in
large measure to the fact that no compelling evidence had been found
for the existence of neutron stars or black holes or for the binary
nature of either of the best known sources, Sco X-1 and Cyg X-2.
When the pulsar in the Crab Nebula was discovered, it became clear
that emission from this neutron star was drawing energy from the
slowing down process described previously and models invoking pulsar mechanisms for all X-ray sources became widely accepted. It
was not until clear evidence was provided from UHURU that Cen X-3
and Her X-1 were, in fact, short period ($\sim$ 1 day) and therefore
close binary systems that serious consideration was given to models
based on accretion onto compact objects in mass transfer binaries.

Since then a large body of observational evidence has been
obtained in X-rays, visible and radio to support this picture. Before proceeding with the description of some of the observational evidence, it might be useful to describe more fully
how the accretion process would actually take place.

Let us consider two stars in a binary system: one of normal
size (such as our own Sun, radius $\sim 10^{11}$ cm), and one condensed.
For a neutron star of 1 $M_\odot$ the accepted radius is of about 10^6 cm.
If we draw the equipotentials about each star, as in Fig. 7, we
find that if the stars move in circular orbits and co-rotate with

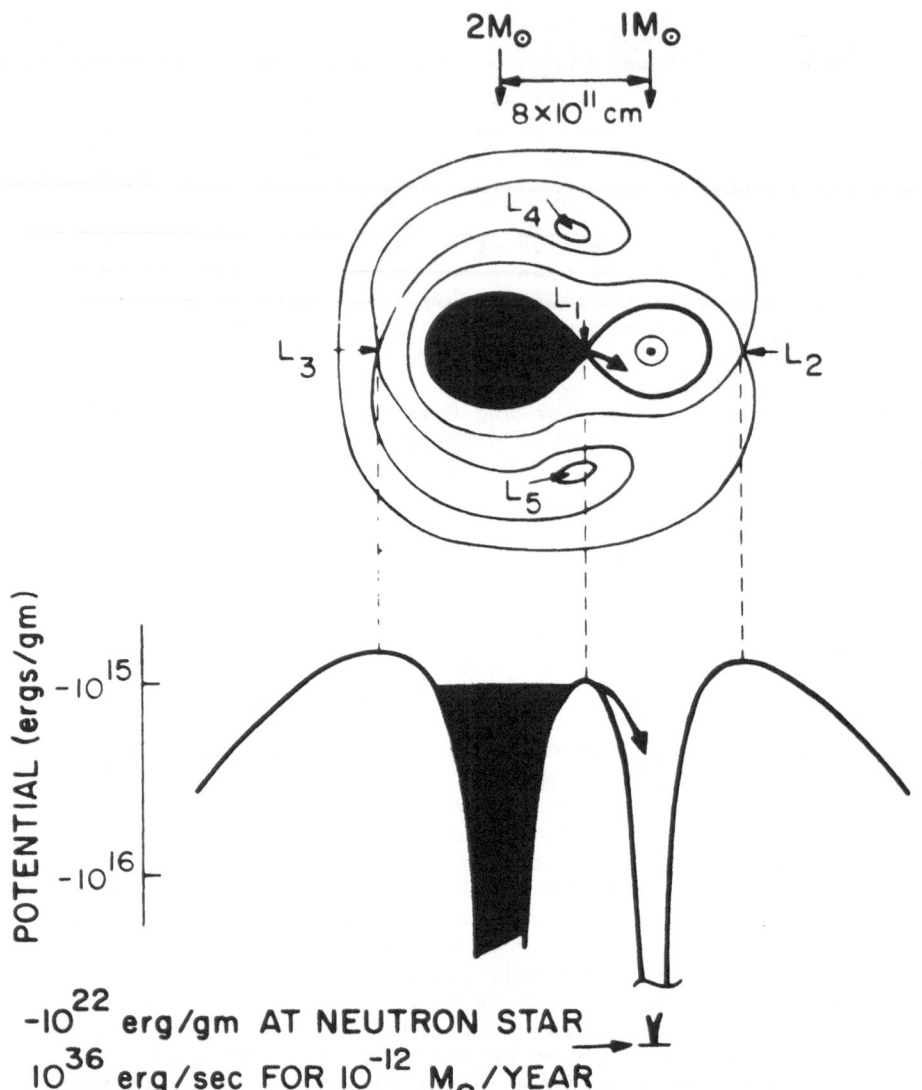

Fig. 7. Equipotential lines of a close binary system formed by a normal star and a neutron star which move in circular orbits and co-rotate with the system.

the system, they are approximately circles of increasing radius about each star. We can continue drawing equipotentials which will be deformed with increasing distance from the stars until the two sets meet. At that point, we have reached a well-known figure 8 configuration, which is often referred to as the Roche lobes, from the French astronomer who first realized their significance. If matter is released at rest inside one lobe, it will fall on the

star within it. Let us suppose that the atmosphere of one star
expands until it overflows its lobe. Then, this matter can fall
onto the other star and, in the case of compact companion, acquire
large amounts of kinetic energy while falling into its deep gravitational well. While, in the case of at least one system (Her X-1),
this overflow of the Roche lobe appears to be actually taking place,
in other X-ray binaries we believe that the material is blown off
one of the stars by means of a stellar wind driven by radiation
pressure. This distinction is important in terms of what is actually observed in X-ray sources, and in terms of the evolution of
the binary systems associated with them; however, the specific origin of the gas flow has little effect on its behavior in the neighborhood of the accreting object.

If we consider this behavior more in detail, we find (as was
pointed out by Prendergast (see Prendergast and Burbidge, 1968))
that because of the angular momentum possessed by the material, it
cannot simply infall radially. It must, therefore, spiral about
the compact star losing angular momentum by viscous interactions

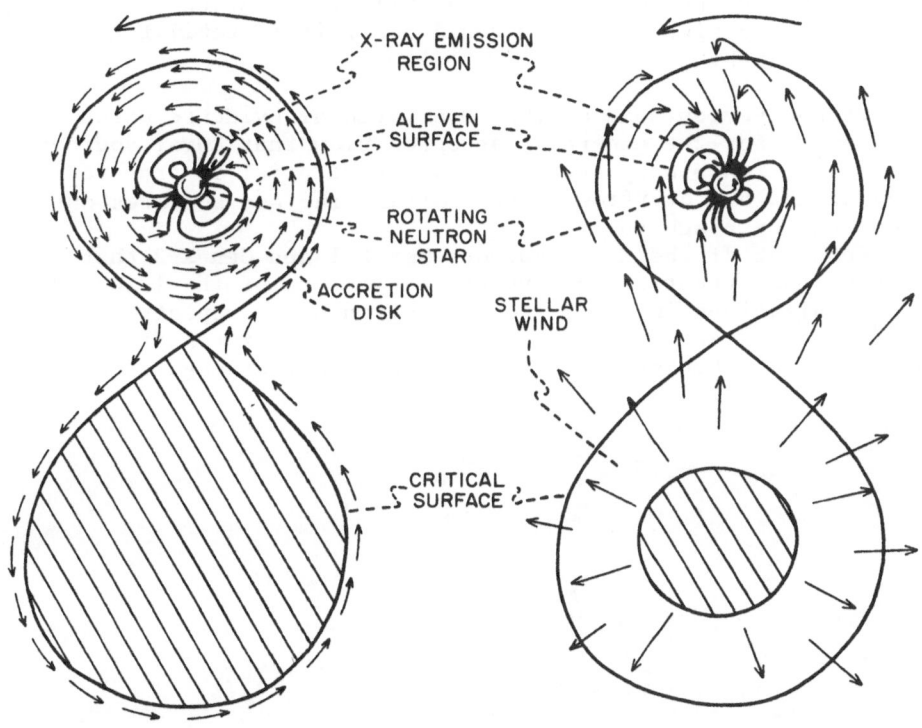

Fig. 8. Schematic diagrams of the flow of matter by Roche lobe
overflow or by stellar wind onto a magnetized neutron star in a
binary system.

with the surrounding gas. It can be shown that the matter will
dispose itself in a differentially rotating thin disk in which a
particle spirals slowly toward the inner edge losing angular momentum until it finally falls onto the compact star. If the compact star is a magnetized neutron star, there will be a radius at
which the magnetic energy density will equal the kinetic energy
density of the accreting gas. Since the gas is exposed to high
energy radiation, it will be ionized and highly conductive. It
will, therefore, be constrained to follow magnetic field lines
and will be guided to the poles of the neutron star (Fig. 8). If
the star is rotating about an axis different from the magnetic
axis, the emitted radiation will appear to us pulsed with a period
equal to the period of rotation of the star. If, on the other
hand, the compact star should be a black hole, then the gas will
disappear from sight as it crosses the Schwartzchild radius of the
object – that being the radius beyond which no radiation can reach
us due to the fact that it will be captured by the star's gravitational field. Thus, the only radiation we will be able to observe would come from the heated gas at the inner edges of the
accretion disk. No periodic pulsations will be observed, although
flickering of very high frequency could occur if instabilities
develop in the disk.

     Let us now review some of the observational evidence which
forms the basis for the generally accepted view that X-ray sources
are associated with binary systems in which material is transferred
either by Roche lobe overflow or by stellar wind and accretes onto
a neutron star or black hole.
     The behavior of the X-ray sources Her X-1 and Cen X-3 that
first caught our attention was the regular nature of the 1.24 and
4.8 second pulsations in the 2 - 6 keV region of the spectrum,
(Fig. 9). Soon after, we noticed that the average intensity seemed
to undergo abrupt changes of as much as one order of magnitude

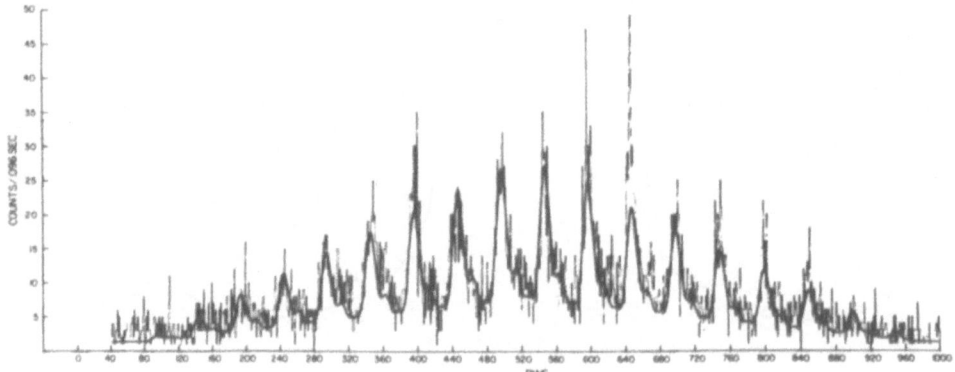

Fig. 9. Counting rate data for Cen X-3 during a 100 sec pass.
The heavier line is a functional fit to the data.

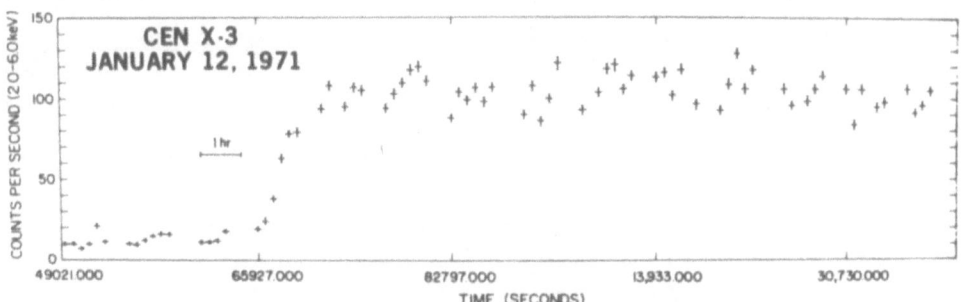

Fig. 10. Explanation in the text.

occurring in an hour (Fig. 10). Observing several such transitions, we were struck by their apparent regularity (Fig. 11), and we interpreted them as due to the periodic orbital motions of the X-ray

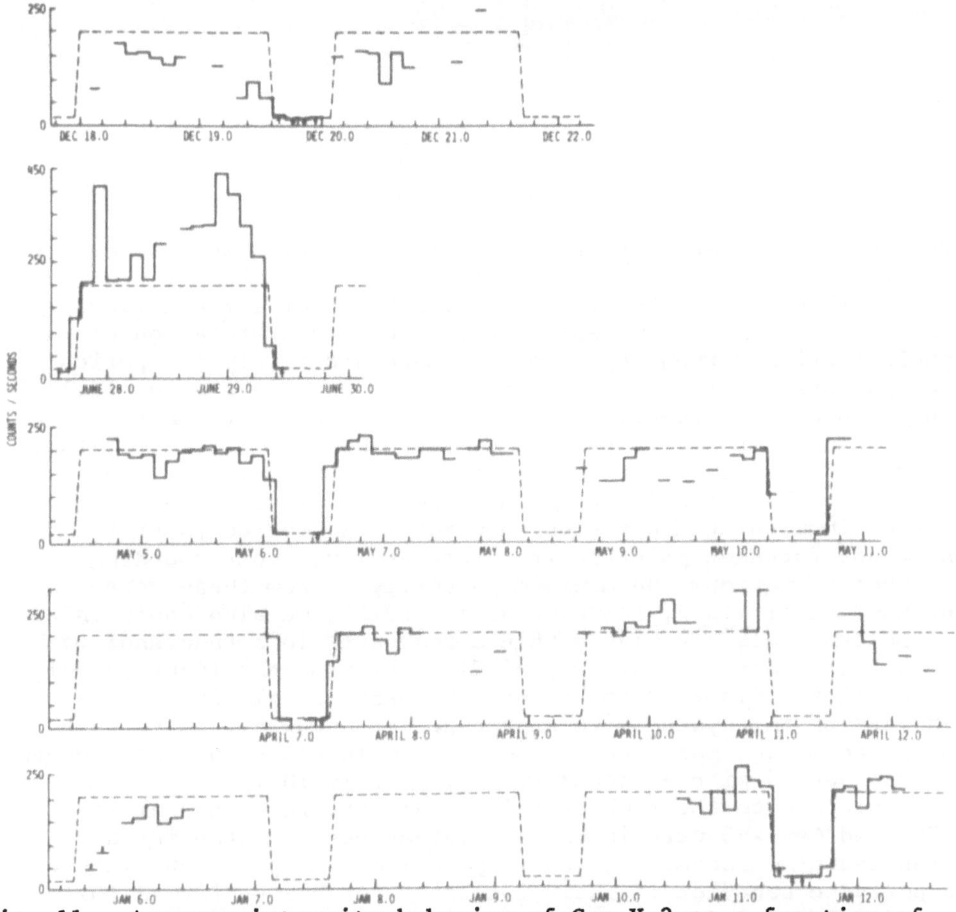

Fig. 11. Average intensity behavior of Cen X-3 as a function of time. See also the text.

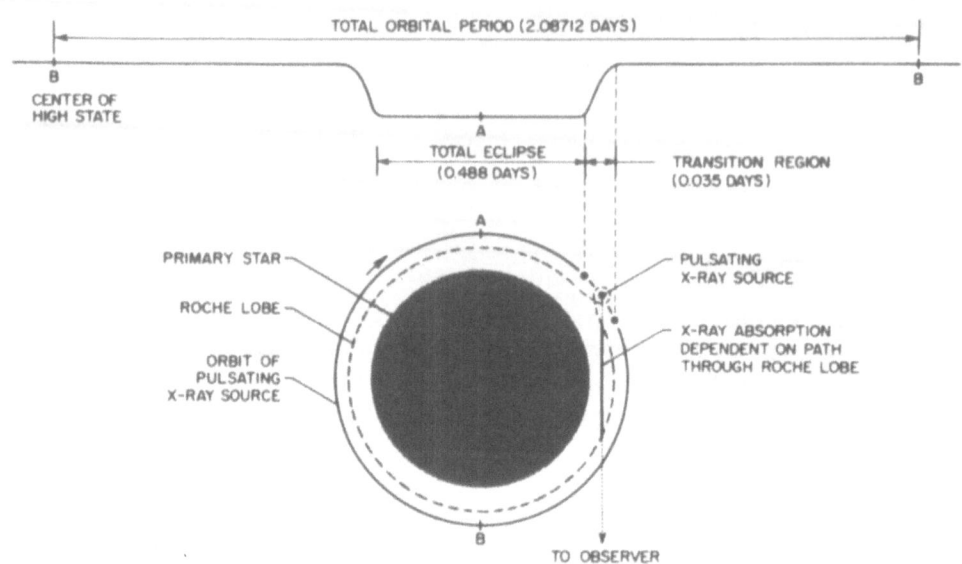

X-RAY BINARY SYSTEM

Fig. 12.   Binary system model for Cen X-3.

source about a companion star. The intensity variation would then
be due to eclipse of the X-ray source once every orbit by the
companion (Fig. 12). The orbital periods were of 1.7 and 2.09 days
for Her X-1 and Cen X-3, respectively. This interpretation was
strongly confirmed when, by examining more accurately the period
of the pulsations, this period was found to vary in a sinusoidal
manner as would be expected from the Doppler shift of the frequency
due to circular orbital motion (Fig. 13). Note in particular that
the phase of the variations agrees exactly with what would be expected under these conditions. The source pulsations appear to
have a minimum change in period when the X-ray source is at the
closest and furthest point in its orbit from us, corresponding to
the center of maximum and minimum intensity. From these data
(Tananbaum et al. 1972; Schreier et al. 1972), we also could infer
that the orbit was circular within a few parts in a thousand; we
could directly measure the orbital diameter from the light travel
line across the diameter and deduce the orbital velocity and reduced mass of the system with unprecedented accuracy. Also, extending the measurement over a week, we could determine the average
pulsation period with an accuracy of 1 part in 10^7.

It was due to these observations that the fact that at least
Her X-1 and Cen X-3 were in binary systems became established.
The subsequent discovery of their optical counterparts showing the
same periodic behavior in their intensity further confirmed this
view and allowed us to unambiguously identify Her X-1 and Cen X-3
with their optical counterparts HZ Her and Kreminsky's star.

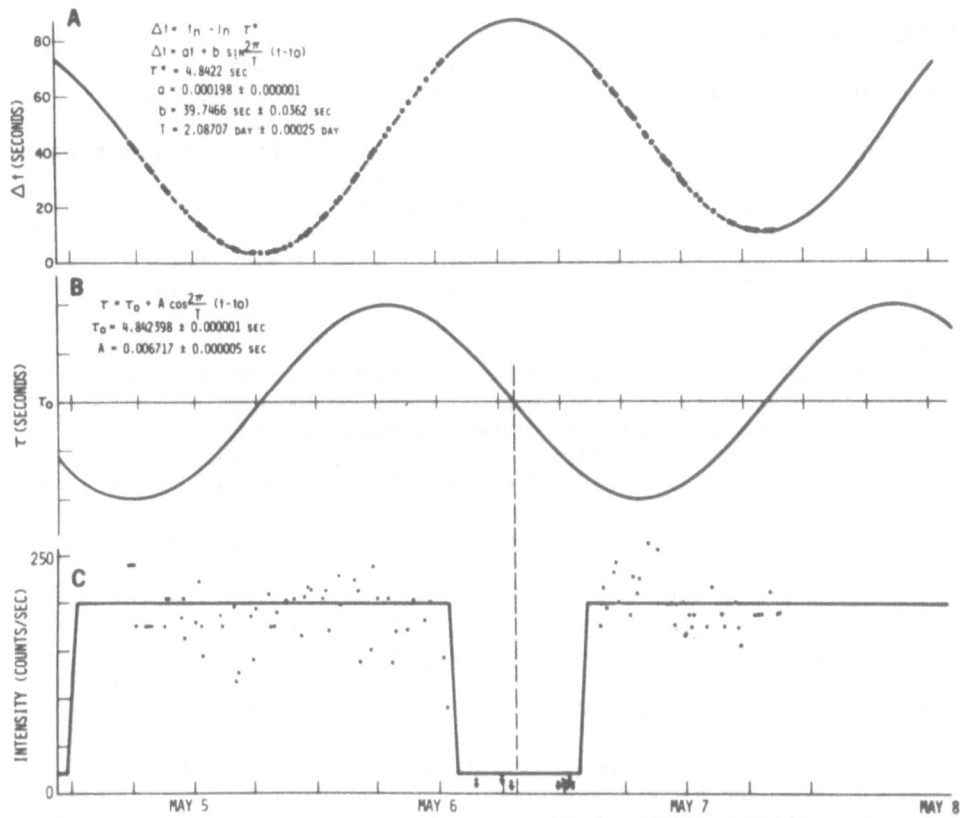

Fig. 13. Bottom: The intensity observed from Cen X-3 (dots) and the light curve predictions for 1971, 5-7 May. Top: The difference Δt between the time of occurrence of a pulse and the time predicted for a constant period, plotted as a function of time. A best fit function and the values of the parameters are given. Center: The dependence of the pulsation period τ on time as derived from the best fit phase function.

Since Her X-1 and Cen X-3 are the only regularly pulsating X-ray sources, the same techniques could not be used in other cases to establish their binary nature. We have relied, therefore, mainly on the variation in average intensity for all those exhibiting eclipses such as 3U1700-37, 3U0900+40, and SMC X-1, and the coincidence between the periods of optical intensity variations and X-ray eclipses. Recently, however, pulsations have been observed from many other sources, including 3U0900+40 and SMC X-1.

In one case, Cyg X-1, the identification with a binary system HDE 226868 was based on positional coincidence between an unusual variable radio source and the optical star, and on a correlated transition occurring in the X-ray and radio regions of the spectrum which established the identification of the radio source with the X-ray source.

As soon as the binary nature of the X-ray source is established, and a visible light companion is identified, we can use
the information on periods and orbital velocities from either or
both wavelength regions to determine the parameters of the system.
It is important to note that the study of orbital motions in binary systems is the only certain method available to us to determine stellar masses. Of crucial importance is also the possibility
of determining the distance of the source, which is normally done
by observing the reddening of the optical companion spectrum due
to interstellar matter. This, then, allows us to determine the
intrinsic luminosity of the X-ray source.

The parameters of the identified binary X-ray sources are
summarized in Table I. It is clear from an inspection of this
Table that most binary sources contain a massive blue supergiant
star (15 - 30 $M_\odot$) and a smaller mass object with masses between
0.7 and 14 $M_\odot$. The companion of Her X-1 is a low mass star ($\sim$ 2 $M_\odot$)
and this might also be the case for Cyg X-3 and for the companions
of Sco X-1 and Cyg X-2, if they should turn out to be binaries as
well. The orbital periods are in the range from 0.2 to 9 days.
The distances are of the order of several kpc and, therefore, the
luminosities are of the order of 10^{36} to 10^{38} erg/sec.

The compact nature of the star on which the X-rays are generated can be inferred by the rapidity of the intensity variations
we observe. In Her X-1 and Cen X-3 we have seen that regular pulsations occur with periods of 1.2 seconds and 4.8 seconds, and in
Cyg X-1 it was discovered that irregular pulsations occur on a
time scale as short as milliseconds. Most other binaries were
observed to exhibit large changes in intensity in times of 0.1 to
1 second. This implies that the region of X-ray emission is
smaller than the distance travelled by light in the same time interval, and must be less than 10^9 cm, or a few times the size of
the earth. We known in astronomy of only three possible states
in the evolution of stars in which matter may reach such high
density that a 1 solar mass could be compressed into these dimensions: white dwarfs, neutron stars, and black holes.

How can we distinguish between these three types of compact
object? The considerations will be of two different kinds: the
first uses arguments based on pulsation periods and stability
similar to those used in the case of the pulsars to discriminate
against white dwarfs; the second relies on the mass determination
and X-ray morphology to distinguish between black holes and neutron stars.

It was soon recognized that the high degree of regularity in
the period of Her X-1 and Cen X-3 implied a stable underlying clock
which could be provided by either stellar rotations or pulsations.
The periods appeared, however, too short to be reconciled with
vibration or rotation of a white dwarf. Thus, we are left with
the choice of either neutron stars, or black holes as the X-ray
source. The existence of regular pulsations in Her X-1 and Cen X-3

seems to decisively rule out black holes, in agreement also with
the low mass estimate. The existence of rapid intensity variations
in X-rays from 0.1 to 0.001 seconds present in all other binaries
seems to also rule out white dwarf candidates for these sources.
In these cases the choice between neutron stars and black holes is
dictated by the mass estimate, and by the morphology of the X-ray
emission. On very general grounds, Ruffini (1973) has proven that
no stable neutron star configuration could be obtained for stellar
masses greater than 3 $M_\odot$. Therefore, compact X-ray sources with
masses larger than 3 $M_\odot$ could be presumed to be black holes (Fig.
14). In addition, the observational appearance of the X-ray emission
should be quite different. No regularly pulsed emission should be
expected from black holes, but rapid flickering could occur with
characteristic times of 10^{-3} to 10^{-5} seconds. For Cyg X-1 the mass
estimate of 8 $M_\odot$ and the existence of rapid flickering have led us
to the conclusion that we are dealing with a black hole.

     The question still arises as to whether the basic energy source
for the X-ray emission could be the loss of rotational energy stored
in the compact object, as is the case in radio pulsars. Even prior
to any measurement, it seemed unlikely that 10^{37} ergs/sec could be
produced by pulsar radiation from such a slowly rotating pulsar.
If this were the case, the lifetime for the system would be quite
short and noticeable increases in the period should have been observable in a few months. The accurate measurement of the average
pulsation period of Her X-1 and Cen X-3 completely rules out this
possibility since it shows (Fig. 15) that the period is not increasing with time, but it is decreasing, so that rotational energy is
being acquired and not lost.

     On the other hand, many of the stars associated with binary
X-ray sources are blue massive supergiants (O and BO) which are
known to lose mass by stellar wind. Thus, a priori, accretion could
be expected to lead to X-ray emission from such binary systems, if
the companion is in fact a compact star.

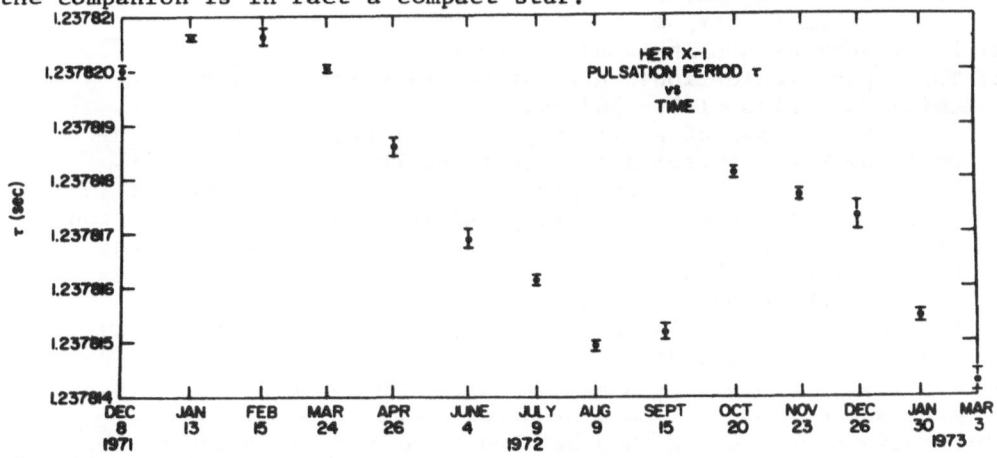

Fig. 14.

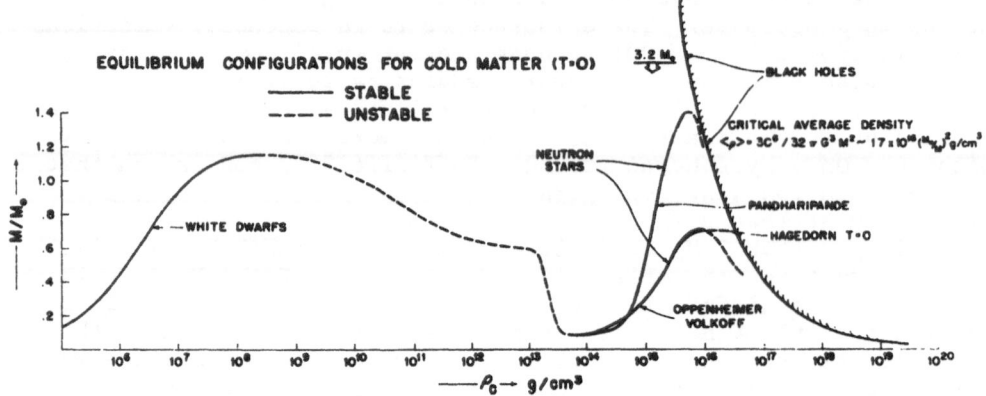

Fig. 15.

There is independent evidence for gas streams and mass transfer in these systems. First, the changes in orbital periods and
in pulsation period can be understood in terms of accretion and
mass loss from the system. Also, visible light spectroscopic
observations give evidence for emission occurring from gas streams
somewhere midway between the two stars. Finally, examination of
the dependence of X-ray emission or orbital phase gives us a direct
measurement of density distribution in the system. Thus, for
instance, in the case of Cen X-3, the slow increase in intensity
over several orbital periods can be beautifully explained in terms
of varying stellar wind density (Giacconi, 1975). The densities
which we measure are precisely those that would be required in an
accretion model to give rise to the expected luminosity, taking
into account the appropriate stellar wind velocities and efficiency
for capture of the material.
      Reviewing the arguments:
a. X-ray sources are in binary systems;
b. The rapid fluctuations in X-ray emission require a compact
   emission region $< 10^6 - 10^9$ cm;
c. Only three types of stars have a radius smaller than 10^9 cm:
   white dwarfs, neutron stars, or black holes;
d. Rotational or vibrational periods in white dwarfs are too long
   to explain the observed X-ray pulsations leaving only neutron
   stars or black holes as candidates for most of the sources;
e. Pulsar-like emission from neutron stars cannot give rise to the
   observed phenomena;
f. Mass streams occur in the binary systems in which the X-ray
   sources are imbedded;
g. The most plausible mechanism is accretion onto neutron stars,
   or black holes, in a mass transfer binary system.
This chain of reasoning then has led to an understanding of the
basic energy source for the compact X-ray sources. However, as

in the case of pulsar emission, we do not fully understand the process by which the gases are heated to high temperature, the details of how the heated plasmas penetrate the magnetosphere of the neutron star and finally the specific process by which X-ray emission occurs. In the next section, I will introduce general considerations about X-ray emission processes and discuss specific sources in which one or the other processes are believed to dominate.

It is interesting in closing to note that the explanation of X-ray emission from most of the compact X-ray sources requires no new physical law. In the case of Cyg X-1, however, we are confronted with a puzzle. If the star is not a black hole and no interpretation can be found that allows us to escape the conclusion that we have a small star ($R < 10^6$ cm) with very large mass ($M \geq 8\ M_\odot$), then we must conclude that new unknown forces can halt the collapse. If, on the other hand, Cyg X-1 is a black hole then a whole new field of physics is opened for our study. In fact, Penrose, Hawking and others are now busily exploring its new laws, presumably taking for granted that such states of matter do exist.

4. PROCESSES FOR X-RAY EMISSION IN COSMIC X-RAY SOURCES

We have in the past few sections explored the question of the energy source for the high luminosity X-ray sources we observe. In a qualitative way we have investigated how release of gravitational energy plays a dominant role in these processes. We have also mentioned possible mechanism of conversion of this energy into heating of plasmas or acceleration of particles. Now I would like to discuss in greater detail the processes resulting in X-ray emission given certain initial conditions. I will again discuss observations of specific objects in this context to illustrate how the study of spectral shape of the observed radiation can give us information on the physical conditions of the source. From this knowledge one can attempt the more arduous step of inferring initial conditions and processes relating the source of energy and its conversion. I will consider four processes which may give rise to X-ray emission in an astrophysical setting:[†]
a. Thermal bremsstrahlung,
b. Synchrotron,
c. Compton scattering,
d. Black body.

[†]I have closely followed the treatment by G. Blumenthal and W. Tucker, shown in Chapter 3 (Mechanisms for the Production of X-rays in a Cosmic Setting) of "X-ray Astronomy", 99-153, R. Giacconi and H. Gursky, co-editors, D. Reidel Publishing Co., 1974, Dordrecht, Holland.

a. Thermal bremsstrahlung

Bremsstrahlung radiation is due to the acceleration of fast electrons in the electric field of atomic nuclei. The cross section for production of a photon of energy $h\nu$ and $h\nu + d(h\nu)$ is given by

$$\frac{d\sigma}{d(h\nu)} = \frac{a\ Z^2 r_e^2 c^2}{v^2}\ \frac{G(E,\ h\ )}{h\nu} \tag{4.1}$$

where a is the fine structure constant, r_e the radius of the electron, Z the atomic number, v and E the velocity and energy of the electron. G, the Gaunt factor, varies slowly as a function of E and $h\nu$, and therefore is taken as constant in an approximate computation. An electron therefore will lose energy at the rate:

$$-\frac{dE}{dt} = \frac{a\ Z^2 r_e^2 c^2}{v}\ NEG \tag{4.2}$$

where N is the number of atoms cm^{-3}. In the case of fast electrons ($E \gg m_e c^2$) not only do we have to consider the interaction of the electrons with the nuclei, but also with the atomic electrons. We therefore substitute for Z^2 the expression $(Z+1)(Z)$. The spectrum of emitted photons is a function of the incident electron spectrum. A special but important case is that of the emission of a hydrogen gas in thermal equilibrium at high temperature. Protons and electrons will therefore have a Maxwellian distribution such that

$$N(v)dv = 4\pi N \left(\frac{m_e}{2\pi kT}\right)^{3/2} v^2\ \exp\left(-\frac{m_e v^2}{2kT}\right) dv \tag{4.3}$$

where T is the temperature, k the Boltzmann's constant and $N = N_e = N_p$ space density of electrons or protons. Neglecting electron-electron collisions, assuming G = constant, we can obtain the power radiated per unit volume and unit energy interval of the protons:

$$\frac{dP_b(T)}{dVd(h\nu)} \approx 10^{-11}\ T^{-1/2}\ N_e^2\ \exp\left(-\frac{h\nu}{kT}\right)\ erg\ cm^{-3}\ s^{-1}\ erg^{-1}. \tag{4.4}$$

If the source is optically thin then the radiation from a volume V will be given by:

$$\frac{dP_b(T)}{d(h\nu)} \approx 10^{-11}\ V\ N_e^2 T^{-1/2}\ \exp\left(-\frac{h\nu}{kT}\right) erg\ s^{-1}\ erg^{-1}. \tag{4.5}$$

$V\ N_e^2$ is often called the emission measure. A complete theory requires us to take into account the composition of the gas, the precise form of G and other processes which may be competing such as line emission and recombination radiation. In Fig. 16, the relative contributions to the total emission for the different processes are shown as a function of temperature. At extremely high temperature, such as we may encounter in the emission of intergalactic gas, the line emission and recombination radiation are

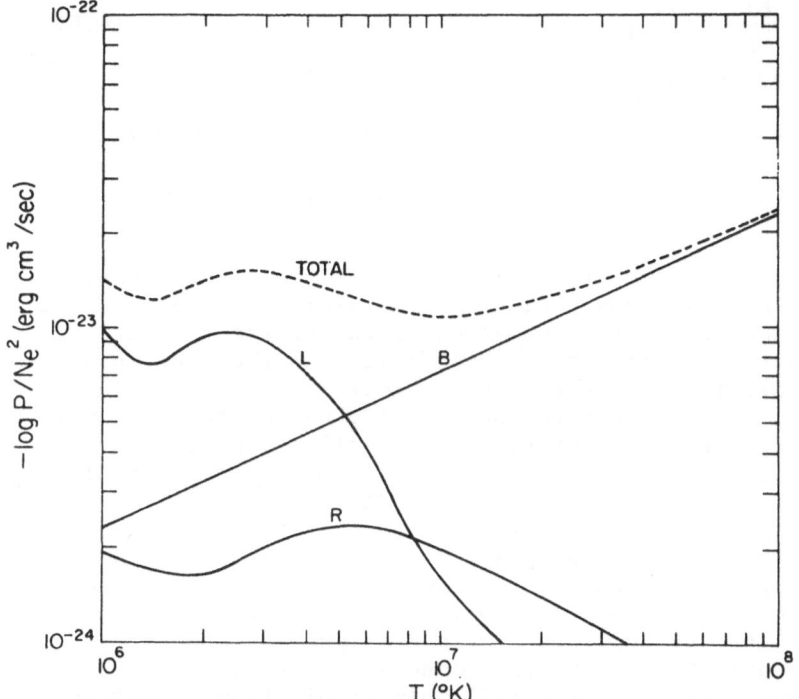

Fig. 16. Total emission from a hot plasma in lines (L), recombination (R), and bremsstrahlung (B) as a function of temperature.

negligible. Even in that case, however, the presence of heavy
elements may change the picture. Certainly line emission and recombination radiation are important in the emission from supernova
remnants and dominate in the case of the Sun.

        Without going into further details on these other effects, I
would like to make some general points.

        The emission by thermal bremsstrahlung is an extremely effi-
cient mechanism of conversion of thermal energy into high energy
photons. High energy electrons in a cold medium radiate only with
an efficiency of 10^{-5}. Most of their energy is lost in ionizing
collisions. It follows then that if we inject into a gas initially
cold electrons with any energy distribution, they will soon expend
their energy in thermalizing the gas which will then efficiently
radiate by thermal bremsstrahlung. This point which was first
made by Rossi in 1964 explains why the emitted spectra from most
of the cosmic sources of this type closely follow the thermal
bremsstrahlung spectrum with essentially an exponential decrease
at high energies.

        An important parameter is the time for cooling of such sources.
In an approximate, but convenient, form:

$$t_c(\text{sec}) \approx \frac{10^{11}}{N_e} T^{1/2} .$$ (4.6)

In two extreme examples for

      X-ray stars $r = 10^{12}$ cm $N_e \sim 10^6$ cm^{-3} $t_c \sim 0.2$s

      Clusters of $r = 1$ Mpc $N_e \sim 10^{-3}$ cm^{-3} $t_c \sim 10^{18}$s
      galaxies

This implies, for instance, that the intergalactic cluster gas, believed to be radiating now could be primordial and have remained at high temperature since the formation of the cluster itself. For small dense regions in the X-ray binaries, on the other hand, a continuous replenishment of the energy lost is required. In practical cases, we are often in one or the other extreme situation. In some cases, such as the steady, soft, uneclipsed source in Cen X-3, while it is attractive to consider thermal bremsstrahlung from a hot optically thin extended source, the problem of the energy input is not resolved. In order to have a long life N_e must be made small but then the emitted power decreases as N_e^2. If N_e is kept large to explain the observed luminosity, then a steady source of heating is required. Only marginally satisfactory agreement is obtained between observations and several present models.

    Another caveat when considering observed X-ray spectra is as follows: An exponential spectrum will only result from an optically thin, isothermal, hydrogen cloud. If several emitting regions are present at different temperatures, almost any spectral shape can be obtained by appropriately choosing the relative contribution of regions of different temperatures.

b. Synchrotron emission

Synchrotron emission is due to deflection of high energy electrons in a magnetic field. A single high energy electron will lose energy at the rate

$$-\frac{dE}{dt} = 1.6 \times 10^{-15} \gamma^2 B^2 \beta^2 \sin^2\alpha \text{ erg s}^{-1}$$ (4.7)

where $\gamma = (1-\beta^2)^{-1/2}$, $\beta = v/c$, α is the pitch angle and B the magnetic field in gauss. Averaging over α, for an isotropic distribution of particles yields

$$\frac{dE}{dt} = 1.1 \times 10^{-15} \gamma^2 B^2 \beta^2 \text{ erg s}^{-1}.$$ (4.8)

For a given β, $dE/dt \sim 1/m^2$, therefore synchrotron radiation from electrons is much greater than for protons; therefore the previous formulas refer to electrons.

    One can show that the maximum in the spectrum occurs at a frequency

$$\nu_m \sim \gamma^2 \ \frac{eB \ \sin \ \alpha}{2\pi \ mc} \tag{4.9}$$

Therefore, if $B < 10^{-4}$ gauss, then $E_e > 10^{14}$ eV. The actual energy differential spectrum goes as $(\nu/\nu_m)^{1/2} \exp{(-\nu/\nu_m)}$ at high energy and as $\nu^{-1/3}$ at low energy.

If we inject electrons of different energies, then the resultant synchrotron spectrum is due to the superposition of individual spectra. For the simple law $n_e(\gamma) \sim \gamma^{-s}$ which is often found in nature, the resultant number spectrum for synchrotron photons of frequency ν is given by $n_\nu \sim \nu^{-(s+1)/2}$. For a power law spectrum of the injected electrons, we obtain a power law spectrum of the emitted photons with a different exponent as indicated above.

The characteristic time for synchrotron energy loss of an electron in terms of the radiated peak frequency ν_m can be rewritten as

$$t_S (\text{sec}) \approx \frac{5 \ x \ 10^{11}}{B^{3/2} \nu_m^{1/2} \ \sin^{3/2} \ \alpha} \tag{4.10}$$

if $\nu \sim 10^{18}$ Hertz typical lifetimes are:

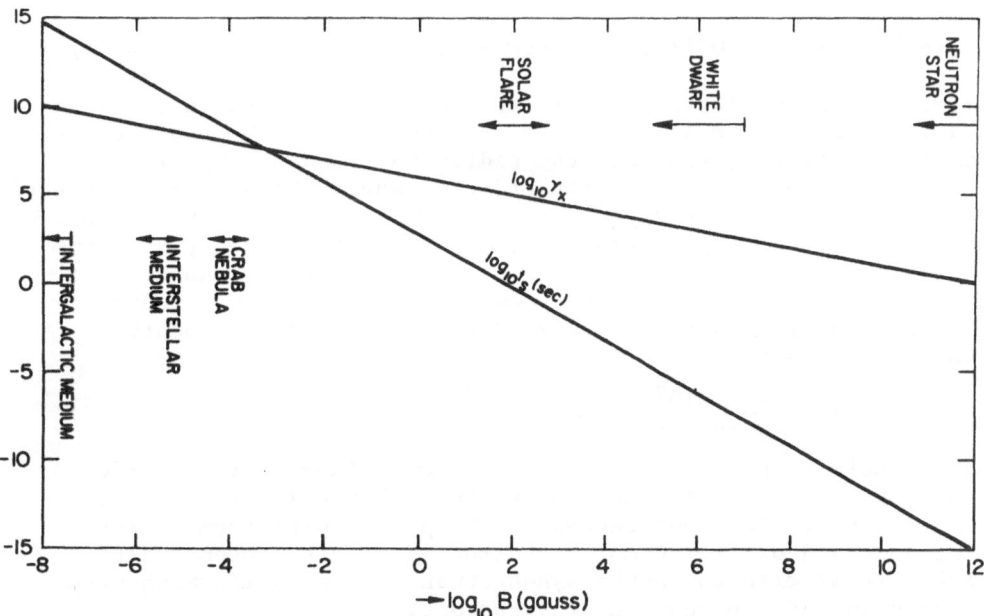

Fig. 17. The synchrotron lifetime and Lorentz factor γ_x for an electron with $\alpha = \pi/2$ emitting synchrotron radiation at X-ray frequencies ($\nu = 10^{18}$ Hz). Also indicated are approximate ranges for the magnetic field in various astronomical regions where X-ray emission may occur.

$$B = 10^{12} \text{ G} \qquad t_s = 10^{-15}\text{s} \qquad \text{(pulsar?)}$$

$$B = 10^{-4} \text{ G} \qquad t_s = 10^{9}\text{s} \qquad .3 - 3 \text{ years in Crab}$$

A graph showing the relation between lifetime and magnetic field
strength is given in Fig. 17. Last, but not least, the radiation
from synchrotron emission should exhibit linear polarization. The
measurement of linear polarization of the X-ray radiation from
Crab obtained by Novick has been generally considered definite
proof of its synchrotron radiation origin.

c. Compton scattering

If u, the total radiation energy density, is greater than the
magnetic field energy density $B^2/8\pi$, then most energy is lost by
electrons in photon collisions rather than in synchrotron losses.

In the Thompson limit ($h\nu_o \ll mc^2$) the collision of a photon
of frequency ν_o with an electron of energy $\gamma m_e c^2$ results in the
average in a photon of frequency $\langle\nu\rangle = 4/3 \ \gamma^2\nu_o$.

When the radiation is isotropic, we can average over angles
to obtain the energy loss per unit time due to collisions with
photons

$$- \frac{dE}{dt} = 2.6 \times 10^{-14} \ \gamma^2 \ u \quad \text{erg s}^{-1}. \tag{4.11}$$

Therefore, in the Thompson limit the energy loss goes as γ^2 as it
does for synchrotron losses, with u instead of B. On the average,
Thompson scattering increase the radiation frequency by $4\gamma^2/3$
in analogy with the synchrotron process, where the emitted frequency
is γ^2 of the cyclotron frequency. By similarity with the synchrotron case, we can expect that if the differential energy spectrum
of electrons obeys a power law with power s, then the number spectrum of the photons will obey a power law with spectral index of
$(s+1)/2$. The lifetime of electrons of energy γmc^2 to Compton
losses is given by

$$t_C(\text{sec}) = \frac{2.1 \times 10^7}{\gamma u} , \tag{4.12}$$

and in Table II has been computed for few values of the incident
frequency and of γ. If the conditions in Table II are correct,
then $t_C < t_S$ in QSO, and therefore the predominant form of X-ray
emission would be from Compton scattering. This presents difficulties if we wish to invoke synchrotron losses as the mechanism
for emission of the lower energy photons.

Caveat on Slopes

We have seen that both the Compton and the synchrotron effects
produce power law spectra for an initial power law electron spectrum. One should be careful, however, in applying these simple

Table II

	ν_o(Hz)	R(cm)	L(erg s^{-1})	u(erg cm^{-3})	γ_e	t_c(s)
Crab	10^{15} (visible)	10^{19}	10^{37}	10^{-12}	30	10^{18}
QSO	10^{15}	$\leq 10^{18}$	10^{47}	$\geq .1$	30	$\leq 10^{7}$
	10^{13} (IR)	$\leq 10^{18}$	10^{47}	≥ 1	3×10^{2}	$\leq 10^{5}$

considerations to real life situations. We have seen that the energy loss goes as γ^2, therefore high energy electrons have shorter lifetimes than low energy electrons. Therefore, if we initially inject a spectrum $n_e(\gamma) \sim \gamma^{-s}$, s will increase at increasing γ with time. If injection is continuous, s goes to s + 1 at high γ's.

For instance, if the loss is due to synchrotron, we can expect a bend in the spectrum at $\gamma_b \approx 10/B_\perp^3 \, t(y)$ where t(y) is the age of the source in years. The photon energy spectrum will reflect such a bend and $\nu^{(1-s)/2}$ will go to $\nu^{-s/2}$ at ν_b(Hz) $\approx 10^9/B_\perp^3 \, t^2(y)$. If there is no injection ν_b is about the same, but the change in spectral index should be much greater. For continuous injection, then we expect the low frequency spectrum (for instance, in radio) to be flatter by 0.5 than the X-ray spectrum. In the Crab we find this difference to be ~ 0.85. One should keep such considerations in mind when attempting to draw from the observed spectral shape conclusions regarding the X-ray emission process.

Low Energy Cut-off

The emitted spectrum of the radiation can be substantially modified by absorption in intervening materials. The most important absorption phenomenon in X-rays is photo-electric absorption by an atom with emission of an electron of energy W = hν − I where I is the ionization potential of the atom. Therefore, hν > > I, for absorption to occur. The photo-electric absorption coefficient $k_p(\nu)$ is proportional to

$\nu^{-8/3}$ if   h$\nu \approx$ I

ν^{-3} if   h$\nu \ll$ I

$\nu^{-7/2}$ if   h$\nu \gg$ I.

This is taken into account by fitting the observed spectrum with the desired spectral shape with a corrective factor of the type

$(E/E_a)^{-8/3}$ which is used to derive the equivalent hydrogen column
density separating us from the source. This is, of course, relatively
meaningless except as an index, because most of the absorption occurs
frequently at the source which presumably has a normal abundance of
elements. In this case the heavier elements do most of the absorption.

d. Black-body

If the source were to be extremely thick to its own radiation, then
we could expect black-body radiation with a spectrum following
Planck's law. To my knowledge, no such instance has yet been observed,
although clearly black-body emission from the surface of a neutron
star must ultimately contribute a small fraction of the total
emission from Her X-1 and similar sources.

I would like now to turn to the application of some of these
considerations to a few of the known X-ray sources. Let us apply,
for instance, these simple considerations to the discussion of the
emission from clusters of galaxies.
The observations show that the Coma Cluster, for instance, contains a large region of X-ray emission with a distribution similar
to that of the galaxies (Gursky et al. 1971). The spectrum shows
no evidence of a cut-off and appears to have an exponential decrease
at high energy (Kellogg et al. 1973). The emission could be either
thermal or non-thermal. If non-thermal, it would be due to synchrotron or inverse Compton.
Let us consider synchrotron emission. It can dominate only
if $B^2/8\pi > u$ where, as a minimum, u is the radiation density of
$3°$ K background radiation of 4×10^{-13} erg cm^{-3}. Thus, $B \geq 3 \times 10^{-6}$
G. In order to obtain X-rays by synchrotron, the electron energy in
such field needs to be of 3×10^{14} eV from the formula (4.9). But
the lifetime t_S for synchrotron losses by these electrons is of
order 6×10^3 years. Therefore, the electrons must be replenished
in that time scale. Since we observe emission over the entire
cluster, of megaparsec dimensions, a central source, such as an
exploding galaxy, cannot furnish these electrons and we must therefore invoke a number of acceleration sites distributed over the
entire cluster, perhaps each of the galaxies. Since we have no
independent evidence on extragalactic cosmic rays, this hypothesis
appears somewhat far-fetched. On the other hand, if $B^2/8\pi < u$,
the Compton losses predominate. Electrons of only 1 G_{eV} ($\gamma = 2 \times 10^3$)
are required to produce the observed X-ray radiation by scattering
on the microwave background with a lifetime of order 10^{17} seconds,
much longer therefore than the lifetime deduced in the assumption
of a synchrotron process previously.
It is to be noted that these low energy electrons of 10^9 eV
would emit copious synchrotron in the range of 5 MHz. Since we
have seen that the Compton and synchrotron spectral indexes must
have the same relation to the index of the incident electron spectrum, then we expect radio and X-ray spectra to have the same slope

which they do within errors. The predicted ratio of fluxes is
also in rough agreement with observations. The observed exponential decrease of the spectrum at high energy could be reconciled
with a Compton spectrum with a break at high energy, but it has
been generally accepted to imply a thermal bremsstrahlung emission.

Fitting the parameters of a thermal bremsstrahlung model to
the X-ray data, one obtains $T = 10^8$ K at a density of 3×10^{-3} cm^{-3}.
This density, when integrated over the cluster yields $M < 1/10\ M_V$,
where M_V is the virial mass of the cluster. The density of the
gas is not sufficient to insure closure, but is comparable to the
mass contained in the galaxies.

Where does this gas come from? Since the time for cooling
is extremely long ($\sim 10^{10}$ years), therefore it could be primordial
gas left behind at the time of galaxy formation. Or, it could be
gas lost by the galaxies and diffusing through the intergalactic medium. A clue to the solution of this problem, as well as a powerful
confirmation of a thermal bremsstrahlung hypothesis, is the recent
detection by Ariel (Culhane, 1975) of a feature in the spectrum which
is interpreted as iron line emission. Such features would be difficult
to reconcile with the Compton model and implies that the material
forming the gas had been processed through nuclear burning in stars.
From the foregoing, it should be clear that some of the simple considerations that can be derived from the study of the details of
the X-ray emission process give important clues to the nature of
the source. I would like, however, to also stress two points:
i)   Clearly all the processes that we have described must be
operating at some level. Our task is not, therefore, to exclude
one or the other, but rather to determine their relative importance;
ii) The foregoing description of the cluster gas emission does not
address itself to the mechanism of heating of the gas. In fact,
several different models have been proposed from ram pressure
heating by gas infalling into the cluster to galactic wakes. To
discriminate among these models, one must relate other measureable
dynamic parameters of the clusters to the X-ray luminosity. The
study of the X-rays alone cannot solve the problem of the energy
source.

For each source which is studied, similar considerations can
and have been applied, and it is surprising how often simple considerations of cooling times, emission measures, etc., can explain
the qualitative features of the sources or exclude some proposed
models. It is particularly fruitful to consider the radiative
losses of the source under study over a broad range of energies
from radio to X-rays because often we can impose much more severe
constraints on the physical parameters such as density of electrons,
energy spectrum and electromagnetic energy density than we could
by study of one wavelength alone. An example of the scope of
phenomena that must be taken into account in this all wavelength
astronomy is, for instance, the model by Grindlay on the emission
from Cen A, a well-known powerful radio source which has now been
studied from radio wavelength to 10^{11} eV γ-rays (Grindlay, 1975).

Cen A is one of the brightest radio sources with huge lobes separated by 5°, inner lobes separated by 4' and a compact microwave structure in the nucleus itself. At 2.2µ the nucleus is
quite bright and at X-ray energy a compact (< 1') variable (∿ days)
X-ray source is observed. High energy emission extends through
the MeV region up to 10^{11} eV as observed by Grindlay (1975).

The measurement in radio, IR and X-rays seems to indicate that
radiation at all these wavelengths has a common source (Fig. 18),
namely a very small region near the nucleus (Schnopper, 1976).
Grindlay takes as his task to explain all of these measurements
and their variation in the simplest possible way. He finds that
he can do so by assuming the existence of two components each
emitting by synchrotron the low energy components of the spectrum,
self-absorbed at high frequency, and producing the high energy
photons by Compton scattering of the same synchrotron electrons
onto the long wavelength photon. He derives parameter for the two
sources, such as angular size (θ_A = 4 x 10^{-4} arc sec and θ_B = 9 x
x 10^{-3} arc sec), magnetic fields (B_A = 2 gauss and B_B = 0.01 gauss),
radiation densities, etc., and constructs a self-consistent model
which takes into account the relative flux densities to be expected
at different wavelengths. the relative slopes, etc., etc. Fig. 19
summarizes the results of his computations. Predictions based on
this model, which, for instance, requires that the hard X-ray flux

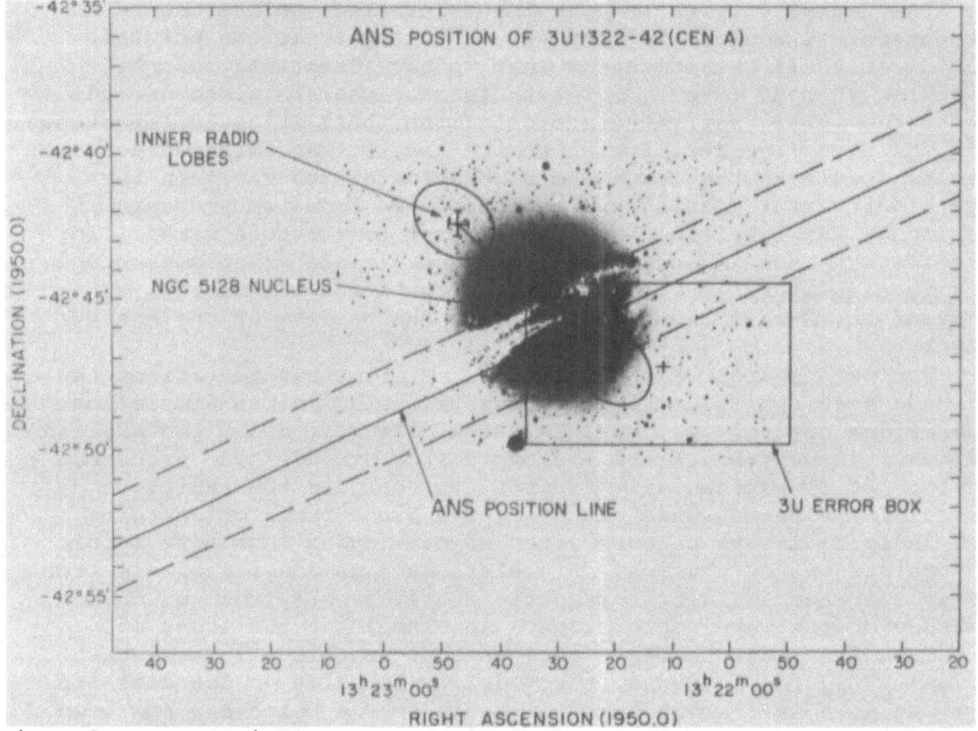

Fig. 18. From Grindlay et al. 1975.

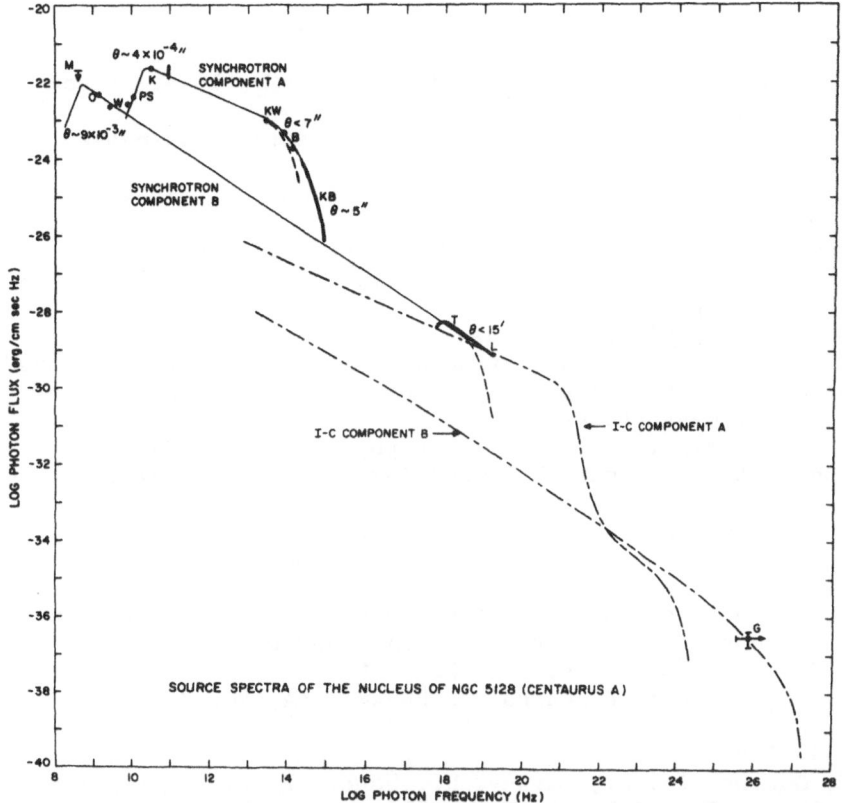

Fig. 19. Observed spectral data and inferred connecting spectra
for the nucleus of NGC 5128. Available data for the source angular
diameter θ are given. The data are well-described by the sum of
two synchrotron spectra (A and B) with assumed cut-offs (dashed
curves) at ~10^{14} Hz and ~3×10^{18} Hz, respectively. The source
component angular diameters are appropriate to synchrotron self-
absorption at ~30 GHz (A) and ~600 MHz (B) and to the calculated
inverse Compton (I-C) spectra from each component (from Grindlay,
1975).

(> 10 keV) varies with the mm flux of component A, and many others,
can be tested experimentally to help us unravel this wonderfully
complex machine. It is precisely in sources such as Cen A that we
expect to find and study objects of 10^7 to 10^9 M_O believed to be
the cause of the explosive events we observe in nuclei of active
galaxies and quasars. By discovering such sources and supplying
additional information and constraints on the physical conditions
of the regions where these high energy events take place, X-ray
astronomy has become a powerful new tool for galactic and extragalactic astronomy.

REFERENCES

Bahcall, J., Bahcall, N., Murray, S. and Schmidt, M., 1975, Astrophys. J. Letters 198, L30.
Blumenthal, G. and Tucker, W., 1974, Ann. Rev. Astron. Astrophys. 12, 23.
Bowyer, C.S., Byram, E.T., Chubb, T.A. and Friedman, H., 1964, Science 146, 912.
Byram, E.T., Chubb, T.A. and Friedman, H., 1966, Science 152, 66.
Catura, R.C. and Acton, L.W., 1975, Astrophys. J. Letters (in press).
Clark, G.W., 1975, CSR = P-75-19.
Clark, G.W., 1976, IAU Circ. No. 2907.
Culhane, J.L., 1976, "Spectra and Structure of Extragalactic Sources", paper presented at the High Energy Astrophysics Division meeting of the AAS, Cambridge, Massachusetts.
Davidson, K. and Ostriker, J., 1973, Astrophys. J. 179, 585.
Elvis, M., Cooke, B.A., Pounds, K.A. and Turner, M.J.L., 1975, Nature 257, 33.
Forman, W., Jones, C. and Tananbaum, H., 1976, Center for Astrophysics Preprint Series No. 457; submitted to Astrophys. J. Letters.
Giacconi, R., 1975a, Center for Astrophysics Preprint Series No. 304.
Giacconi, R., 1975b, Proceedings of the Seventh Texas Symposium on Relativistic Astrophysics, P.G. Berman, E.J. Fenyes and L. Motz, editors, New York Academy of Science, 262, 299.
Giacconi, R., 1976, American Journal of Physics 44, No. 2.
Giacconi, R., Gursky, H., Paolini, F.R. and Rossi, B.B., 1962, Phys. Rev. Letters 9, 439.
Giacconi, R., Gorenstein, P., Gursky, H., Usher, P.D., Waters, J.R., Sandage, A., Osmer, P. and Peach, J.V., 1967, Astrophys. J. Letters 148, L129.
Giacconi, R., Murray, S., Gursky, H., Kellogg, E., Schreier, E., Matilsky, T., Koch, D. and Tananbaum, H., 1974, Astrophys. J. Suppl. Series, No. 237, 27, 37-64.
Giacconi, R. and Gursky, H. (co-editors), X-ray Astronomy, 1974, D. Reidel Publishing Co., Dordrecht, Holland, pp. 155-168.
Gorenstein, P. and Tucker, W., 1975, Center for Astrophysics Preprint Series No. 418; submitted to Ann Rev. Astron. Astrophys.
Grindlay, J.E., 1975, 1975, Astrophys. J. 199, 49.
Grindlay, J.E., Schnopper, H., Schreier, E.J., Gursky, H. and Parsignault, D., 1975, Astrophys. J. Letters 201, L133.
Grindlay, J.E., Gursky, H., Schnopper, H., Parsignault, D., Heise, J., Brinkman, A.C. and Schrijver, J., 1976, Center for Astrophysics Preprint Series No. 455; submitted to Astrophys. J. Letters.
Gursky, H., Kellogg, E., Murray, S., Leong, C., Tananbaum, H. and Giacconi, R., 1971, Astrophys. J. Letters 167, L81.
Gursky, H. and Schreier, E., 1975, Variable Stars and Stellar Evolution - Proceedings of the IAU Symposium No. 67, Moscow, Sherwood and Plaut, co-editors, pp. 413-464.
Hearn, L.R. and Richardson, J.A., 1975, IAU Circ. No. 2890.
Holt, S.S., 1975, GSFC Preprint X-661-75-232.

Jones, C. and Forman, W., 1976, IAU Circ. No. 2913.
Kellogg, E., Murray, S., Giacconi, R., Tananbaum, H. and Gursky, H., 1973, Astrophys. J. 185, L13.
Kellogg, E., 1974, X-ray Astronomy, R. Giacconi and H. Gursky, co-editors, D. Reidel Publishing Co., Dordrecht, Holland, pp. 321-357.
Lamb, F.K., 1974, "Theories of Binary X-ray Sources", Invited paper at International Conference on X-rays in Space.
Margon, B. and Ostriker, J.P., 1973, Astrophys. J. 186, 91.
Mewe, R. et al., 1975, paper presented at COSPAR/IAU Symposium on Fast Transients in X- and Gamma-rays, Varna, Bulgaria.
Prendergast, K. and Burbidge, G., 1968, Astrophys. J. 151, L83.
Rappaport, S., Buff, J., Clark, G., Matilsky, T. and McClintock, J., 1976, CSR-P-76-4; submitted to Astrophys. J. Letters.
Ricketts, M.J., Cooke, B.A. and Pounds, K.A., 1975, preprint.
Rossi, B.B., 1974, Astronomia in Raggi X, Academia Nazionale dei Lincei, Rome, Italy.
Rowan-Robinson, M. and Fabian, A.C., 1975, Mon. Not. Roy. Astron. Soc. 170, 199.
Ruffini, R., 1973, Black Holes, C. and B.S. DeWitt, co-editors, Gordon and Breach, New York, New York, pp. 451-496.
Sandage, W.R., Osmer, P., Giacconi, R., Gorenstein, P., Gursky, H., Waters, J.R., Bradt, H., Garmire, G., Sreekantan, B.V., Oda, M., Osawa, K. and Jugaku, J., 1966, Astrophys. J. 146, 316.
Schreier, E., Levinson, R., Gursky, H., Kellogg, E., Tananbaum, H. and Giacconi, R., 1972, Astrophys. J. Letters 172, L79.
Schreier, E., Schnopper, H., Gursky, H. and Parsignault, D., 1975, Center for Astrophysics Preprint Series No. 333; submitted to Astrophys. J. Letters.
Schnopper, H., 1976, "Extragalactic Compact X-ray Sources", paper presented at the High Energy Astrophysics Division meeting of the AAS, Cambridge, Massachusetts.
Shklovksy, I., 1967, Astrophys. J. Letters 148, L1.
Tananbaum, H., Gursky, H., Kellogg, E., Levinson, R., Schreier, E. and Giacconi, R., 1972, Astrophys. J. Letters 174, L143.
Tananbaum, H. and Hutchings, J.B., 1975, Proceedings of the Seventh Texas Symposium on Relativistic Astrophysics, P.G. Berman, E.J. Fenyves, and L. Motz, editors, New York Academy of Science 262, 299.
Ulmer, M.P. and Murray, S., 1976, preprint.
Willmore, A.P., 1976, preprint.

SUBJECT INDEX

G. Setti (ed.), The Physics of Non-Thermal Radio Sources, 271–287. All Rights Reserved.
Copyright © 1976 by D. Reidel Publishing Company, Dordrecht-Holland.